Revised Research Plan
for the U.S. Climate Change
Science Program

A Report by the
Climate Change Science Program and
the Subcommittee on Global Change Research

May 2008

Table of Contents

Executive
Summary

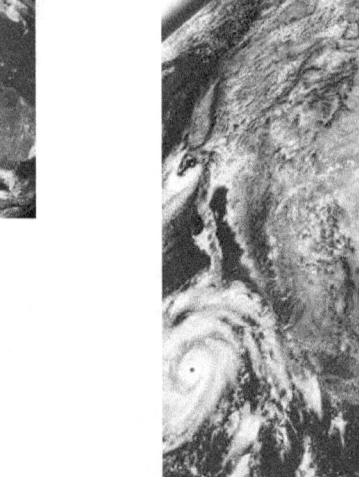

Executive Summary

The U.S. Climate Change Science Program (CCSP) released its Strategic Plan in 2003. This Revised Research Plan, in compliance with Section 104(a) of the Global Change Research Act of 1990, is an update to the 2003 Strategic Plan. It reflects both scientific advances since the publication of the 2003 Strategic Plan and the evolving needs of society. The update focuses on near-term (1-3 year) planning needs, and specifically addresses research plans for the period 2008 to 2010. The Revised Research Plan also represents one of the first steps in the longer term development of the next Strategic Plan. CCSP is currently developing a process for gathering input from a wide range of stakeholder and scientific communities to inform the development of that new Strategic Plan, which will describe approaches for addressing the Nation's needs for climate change information beyond the 2010 time frame and into the next decade.

The Revised Research Plan contains an updated statement of capabilities and objectives consistent with CCSP's current Strategic Plan but reflecting both scientific progress and the Nation's evolving societal and environmental needs. It also contains examples of research progress and a discussion of the program's emerging priorities. Using the program's five strategic goals as an organizing framework, the Revised Research Plan provides a goal-by-goal overview of emerging research questions and themes, key research topics, and illustrative research plans for the years 2008 to 2010.

CCSP's vision is *a Nation and the global community empowered with the science-based knowledge to manage the risks and opportunities of change in the climate and related environmental systems. Its mission is to facilitate the creation and application of knowledge of the Earth's global environment through research, observations, decision support, and communication.*

This mission arises from the recognition that climate variability and change will continue to influence society directly and indirectly, and that in order to make informed decisions, society requires knowledge as to 1) what is changing and how; 2) what forces are causing those changes; 3) how the Earth system may change in the future and affect societies and ecosystems; 4) what parts of the Earth system are most sensitive to global change and how adaptable those parts are; and 5) how scientific knowledge can be effectively applied to manage the risks and opportunities. These are the essence of CCSP's strategic goals.

To address these goals, CCSP will utilize the Nation's investments and advances in monitoring, analysis, and modeling to increase understanding of Earth systems and the processes affecting their changes (Goals 1 through 3), while strengthening activity in understanding societal and ecosystem sensitivities and adaptability, and supporting decisionmaking (Goals 4 and 5). The effectiveness of this suite of activities will be enhanced through the development of robust communication and education activities that serve decisionmakers, stakeholders, and the public.

For each of CCSP's strategic goals, the Revised Research Plan provides a series of specific examples of research priorities and plans. **The illustrative examples in the Revised Research Plan are not intended to**

provide an exhaustive list of every CCSP research project, but rather to provide an indication of the breadth and depth of the program's activities. In addition to the illustrative research examples identified in the Revised Research Plan, it is fully expected that other important research topics, yet to be determined, will emerge from future scientific progress, events, and societal needs.

As identified in the Revised Research Plan, key components of CCSP's activities over the next 3 years include the following:

- CCSP will continue to provide the basic physical science required to understand Earth's past and present climate, including its natural variability, and to improve understanding of the causes of and uncertainties in observed variability and change at global, continental, regional, and local scales. CCSP remains committed to basic, ongoing research to understand climate processes and the forcing factors that cause changes in climate and related systems.

- CCSP will increasingly address emerging needs for research to more fully understand the impacts of climate change on unmanaged and managed ecosystems, human health and infrastructure, economic, and other human systems.

- CCSP will continue to generate science to support decisionmaking related to the management of risks and opportunities of climate variability and change, including adaptive management and mitigation efforts, with an increased emphasis on generating scientific results at regional and local scales.

- CCSP will place greater emphasis on communicating with users and stakeholders (e.g., state and local governments, academia, industry, public utilities, and nongovernmental organizations), both to gain the benefit of their experience, perspectives, and input and to ensure that the results of CCSP research, monitoring data, and assessments are widely and easily available and accessible to potential users of this information.

These four points distill the key similarities and differences between the 2003 Strategic Plan and the way forward that is identified and illustrated in the Revised Research Plan.

CCSP's role in meeting the challenges of global change is vital. CCSP adds value to Federal agency efforts in climate change research and related activities by providing a structure and coordination mechanism that leverages individual agency efforts through increased cooperation, collaboration, and the joint development of research priorities. The CCSP framework allows Federal agencies engaged in global change research to do more than individual agencies could do separately, thus to more effectively address the Nation's global change science needs and coordinate and communicate their activities with those of their domestic and international research partners and stakeholders. In contributing to this framework, the Revised Research Plan is consistent with the CCSP guiding vision identified above and restated here: "A Nation and the global community empowered with the science-based knowledge to manage the risks and opportunities of change in the climate and related environmental systems."

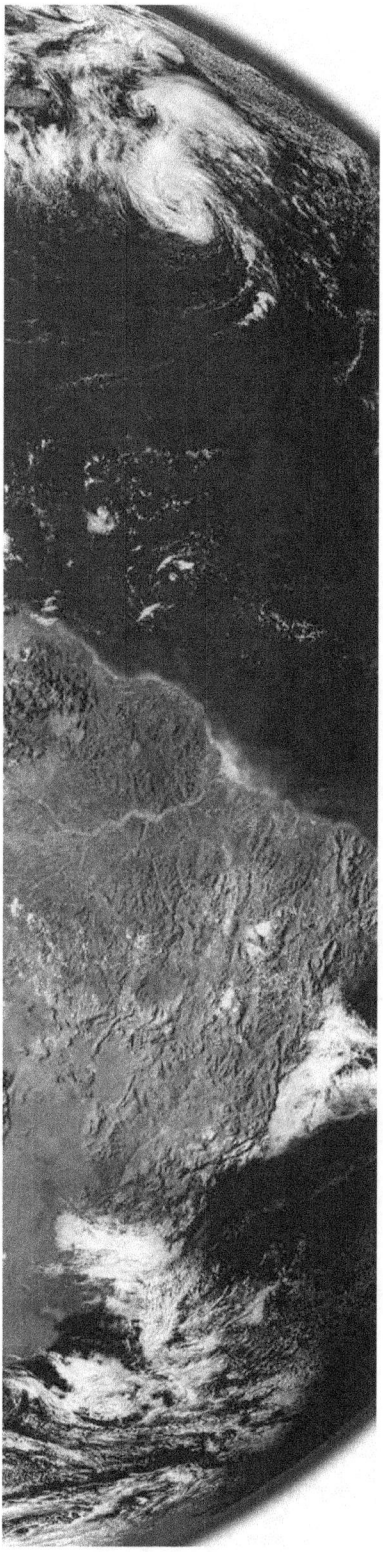

I Introduction

I. Introduction

About the Revised Research Plan

This Revised Research Plan is an update to the 2003 *Strategic Plan for the U.S. Climate Change Science Program* (CCSP).[1] The 2003 CCSP Strategic Plan is an in-depth document, the result of extensive study and stakeholder involvement, intended to provide guidance to and goals for the program for the 10-year period 2003 to 2013 (CCSP, 2003). As such, it provides the larger framework for the activities of the program, and operates at a strategic level as well as providing concrete milestones to guide scientific research and for use in assessing progress toward the overarching goals of the program.

The 2003 CCSP Strategic Plan was developed via a thorough, open, and transparent multi-year process involving a wide range of scientists, managers, and stakeholders, including members of the academic community, Federal, state, and local agencies and other entities, and citizens. A significant part of this process was the review of both the draft and final plan by the National Research Council.[2] These reviews played an important role in influencing the 2003 Strategic Plan's development and its subsequent use.

The Strategic Plan is of long-term value to CCSP and was written for the period 2003 to 2013, but like any strategic plan, it must be supplemented by shorter term revisions that take into account both advances in the science and changes in societal and environmental needs. CCSP has a robust, ongoing long-range strategic planning process to ensure that these needs are met. This Revised Research Plan draws on CCSP's long-range planning process and provides this update. The publication of this Revised Research Plan is also one of the first steps in the longer term development of the next CCSP Strategic Plan.

At the time of this writing, CCSP is engaged in the production of four documents reflecting its current short- and long-term planning and assessment activities in addition to the synthesis and assessment products currently in production.

CCSP Revised Research Plan	May 2008
Scientific Assessment	May 2008
'Capstone' Synthesis and Assessment Product	2009
CCSP Strategic Plan 2013 to 2023	2010

This Revised Research Plan and its companion Scientific Assessment are being produced in compliance with the requirements of the Global Change Research Act of 1990. The Revised Research Plan and the Scientific Assessment will be published on or before 30 May 2008. The Revised Research Plan by definition is intended to address near-term (1-3 year) needs, thus focuses on planning and activities for the period 2008 to 2010, including those already in progress at publication. It draws from and synthesizes existing and new research plans and priorities that are developed and published each year in CCSP's annual report to Congress. It thus provides an integrated, short-range update to the 2003 CCSP Strategic Plan but does not replace that Plan. The specific objectives of the Scientific Assessment – entitled *Scientific Assessment of the Effects of Global Change on the United States* – are to integrate, evaluate, and interpret the findings of CCSP to support informed discussion of the relevant issues by decisionmakers, stakeholders, the media, and the general public. The Scientific Assessment draws from and synthesizes findings from recently published assessments of the science, including reports and products by the Intergovernmental Panel on Climate Change (IPCC) and CCSP.

In addition to producing these two deliverables, CCSP is engaged in developing a long-range strategic planning exercise with the goal of producing the next CCSP Strategic Plan, in order to address the Nation's need for climate change information beyond the short term and extending decades into the future. While the strategic planning process is still in the early stages of development, it will be a thorough, deliberative process that will include intensive stakeholder engagement and multiple opportunities for stakeholders to provide input. It is anticipated that these activities will include such venues as sessions at professional meetings and/or workshops, and that these opportunities will be posted well in advance on the CCSP web site.[3] The new Strategic Plan will also consider and benefit from the comments and discussion generated during the production of the Revised Research Plan and Scientific Assessment.

---

[1] See <www.climatescience.gov/Library/stratplan2003/final>.
[2] See <www.nap.edu/catalog.php?record_id=11565> for the draft plan and <www.nap.edu/catalog.php?record_id=10635> for the final plan.
[3] See <www.climatescience.gov>.

The fourth product in development is a 'Capstone' to the series of 21 CCSP synthesis and assessment products that are being released during the 2006 to 2008 time period. This Capstone Synthesis and Assessment Product will incorporate findings from the suite of 21 synthesis and assessment products, thus will integrate and analyze CCSP results across the full range of sectors, disciplines, and cross-cutting issues treated in the synthesis and assessment products, with a particular emphasis on North America.

Subsequent sections of this Revised Research Plan include: 1) an updated statement of vision, goals, and capabilities consistent with CCSP's current Strategic Plan but reflecting both scientific progress and the evolution of the program based on accomplishments and evolving societal and environmental needs; 2) highlights of ways in which the program is evolving in the context of the progress made over the years since the 2003 Strategic Plan was put in place, and a description of recent priorities that have emerged as a result; and 3) at an integrative level, a description of research plans for the coming years, in order to build upon the work envisioned in the Strategic Plan and begun over the past 4 years.

About the Climate Change Science Program

The vision of CCSP is:

> *A Nation and the global community empowered with the science-based knowledge to manage the risks and opportunities of change in the climate and related environmental systems.*

The core precept that motivates CCSP is that the best possible scientific knowledge should be the foundation for the information required to manage climate variability and change and related aspects of global change. Thus, the mission of CCSP is to:

> *Facilitate the creation and application of knowledge of the Earth's global environment through research, observations, decision support, and communication.*

CCSP's five strategic goals are:

- CCSP Goal 1: Improve knowledge of the Earth's past and present climate and environment, including its natural variability, and improve understanding of the causes of observed variability and change.

- CCSP Goal 2: Improve quantification of the forces bringing about changes in the Earth's climate and related systems.

- CCSP Goal 3: Reduce uncertainty in projections of how the Earth's climate and related systems may change in the future.

- CCSP Goal 4: Understand the sensitivity and adaptability of different natural and managed ecosystems and human systems to climate and related global changes.

- CCSP Goal 5: Explore the uses and identify the limits of evolving knowledge to manage risks and opportunities related to climate variability and change.

These strategic goals provide the overarching framework for CCSP research, and serve to implement and augment the research elements outlined in the Global Change Research Act of 1990 (see Appendix 3 for further explanation of the relationship between the GCRA research elements and the CCSP strategic goals).

In working toward these strategic goals, CCSP employs four core approaches, including:

- Scientific Research: Plan, sponsor, and conduct research on changes in climate and related systems.

- Observations: Enhance observations and data management systems to generate a comprehensive set of variables needed for climate-related research.

- Decision Support: Develop improved science-based resources to aid decisionmaking.

- Communications: Communicate results to domestic and international scientific and stakeholder communities, stressing openness and transparency.

The first two of these core approaches are comparatively mature, drawing upon the scientific strengths of Federal and academic institutions and the combination of valuable legacy data sets with current observational, experimental, and modeling capabilities and future platforms, networks, and data systems. The 2003 Strategic Plan notes that the latter two – Decision Support and Communications – are areas where substantial growth is required, and for which the development of new capabilities and activities will be needed. As new scientific information has become available, it is clear that CCSP is moving from an environment in which understanding the climate system is the focus to one in which the focus is balanced among understanding the climate system and understanding and communicating the effects of changing climate on the environment and society. This transition requires CCSP to engage with a larger number of more diverse consumers of CCSP's science, and includes the need for dialog and partnership with a larger range of agencies, observational systems, and cultures than CCSP has engaged in the past. While substantial progress has been made in these areas over the past few years, decision support and communications remain areas of both need and opportunity for CCSP and for the entire global change community, as pointed out in the 2007 National Research Council (NRC) report on CCSP (NRC, 2007a).

In order to understand CCSP's role in fostering and coordinating Federally funded U.S. climate change research, it is important to understand what CCSP is and the role CCSP has in the Federal government. CCSP has responsibility for implementing some provisions of the Global Change Research Act of 1990, but CCSP is not a Federal agency. Rather, it is a structure and a mechanism for coordinating and integrating Federal research on climate and global change, and for making recommendations on priorities that Federal agencies consider in their planning, as authorized in the Global Change Research Act of 1990. Research on global change, including climate change, is sponsored and/or conducted by 13 Federal agencies and/or departments that are members of CCSP; CCSP is open to all agencies that have an interest in climate change research, including impacts and/or the adaptation to and mitigation of climate change effects. The CCSP membership also includes government entities that do not sponsor or conduct research but which play a critical role in the Federal research establishment. The latter are the Office of Science and Technology Policy, the Council on Environmental Quality, and the Office of Management and Budget. CCSP encompasses both the U.S. Global Change Research Program (established by the GCRA) and the 2001 Climate Change Research Initiative.

CCSP fosters coordination of Federal global change activities across thematic and cross-cutting elements that utilize the four core approaches (scientific research, observation, communication, and decision support) highlighted above. CCSP also helps to coordinate international research and cooperation. Member agencies and departments include the following:

- Department of Agriculture
- Department of Commerce
- Department of Defense
- Department of Energy
- Department of Health and Human Services
- Department of the Interior
- Department of State
- Department of Transportation
- Agency for International Development
- Environmental Protection Agency
- National Aeronautics and Space Administration
- National Science Foundation
- Smithsonian Institution.

The program is led by an interagency committee of senior representatives from the participating departments and agencies that is responsible for overall priority-setting, program direction, and management review. The CCSP Director chairs this committee. Interagency Working Groups, composed of designated Federal program managers representing agencies involved in each of the program's research and cross-cutting elements, work together to plan and implement interagency activities and priorities aligned with CCSP's goals. CCSP's research elements include:

- Atmospheric Composition
- Climate Variability and Change / Modeling
- Global Water Cycle
- Ecosystems

  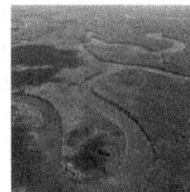  

  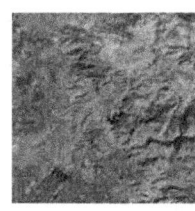  

- Land-Use and Land-Cover Change
- Global Carbon Cycle
- Human Contributions and Responses.

CCSP's cross-cutting elements include:

- Observations and Data Management
- Modeling
- Decision Support
- International Research and Cooperation
- Communication.

CCSP has a single office, the function of which is to facilitate and support the activities of the Program by providing value-added staffing and day-to-day coordination of CCSP-wide program integration, strategic planning, product development, and communication. CCSP's budget process is implemented within the context of the Federal budget process: budget issues are coordinated by the CCSP Principals, CCSP Financial Operations Group, interagency working groups, and other mechanisms, but ultimate accountability for budgetary matters resides with CCSP's participating agencies and departments.

Global change research activities across CCSP's 13 departments and agencies include research conducted by scientists in Federal agencies, academia, industry, and non-profit organizations through a mix of directed and competed programs. The value of CCSP is in the facilitation of interagency cooperation and collaboration; thus, some CCSP accomplishments, while primarily the work of single agencies, have benefited from and been strengthened through the perspectives of and ongoing discussion with other CCSP agencies. Other accomplishments reflect coordinated efforts among multiple agencies that because of their scope or size are activities that could not be undertaken by any single agency. Both types of accomplishments are important. CCSP provides leverage for individual agency efforts in global change research through improved coordination and communication, and provides an avenue for integrating research across agencies, and for producing annual reports to Congress that include both periodic reports on research progress and updated summaries of future research plans. CCSP also provides

climate-related input to other Federal and Administration initiatives (e.g., the Ocean Action Plan, the U.S. Group on Earth Observations), and a way for the Federal climate change research establishment to assess joint opportunities and needs for programmatic evolution in response to changing societal and environmental needs. Scientists supported by CCSP-participating agencies contribute scientific expertise, and CCSP itself contributes other resources to key international efforts, including the United Nations Framework Convention on Climate Change (UNFCCC), the synthesis and reporting activities of the IPCC, the International Council for Science (ICSU), the World Meteorological Organization (WMO), the United Nations Environment Programme (UNEP), and other international partnerships and cooperative activities.

CCSP also supports U.S. participation in the activities of the IPCC. The IPCC is a scientific intergovernmental body established by WMO and UNEP to provide decisionmakers and others with an objective source of information about climate change. The IPCC periodically compiles and releases comprehensive, objective, open, and transparent assessments using the latest scientific, technical, and socioeconomic literature produced worldwide to aid in the understanding of intrinsic climate variability and feedback mechanisms and the risk of human-induced climate change, its observed and projected impacts, and options for adaptation and mitigation. The fourth such analysis, termed the IPCC Fourth Assessment Report, was released in 2007. In addition to providing scientific expertise and leadership by CCSP-participating agency scientists during the drafting and editing process, CCSP directly supported the activities of the Working Group I Technical Support Unit. Via its coordination office, CCSP managed the U.S. author nomination process, the Expert and Government Reviews, and the final government review of the Summaries for Policymakers for Working Groups I and II and the Synthesis Report.

CCSP produces a number of key products. First and foremost, numerous peer-reviewed scientific papers are published each year by scientists working on research that falls under the auspices of CCSP. The importance of these numerous scientific publications is paramount. Without

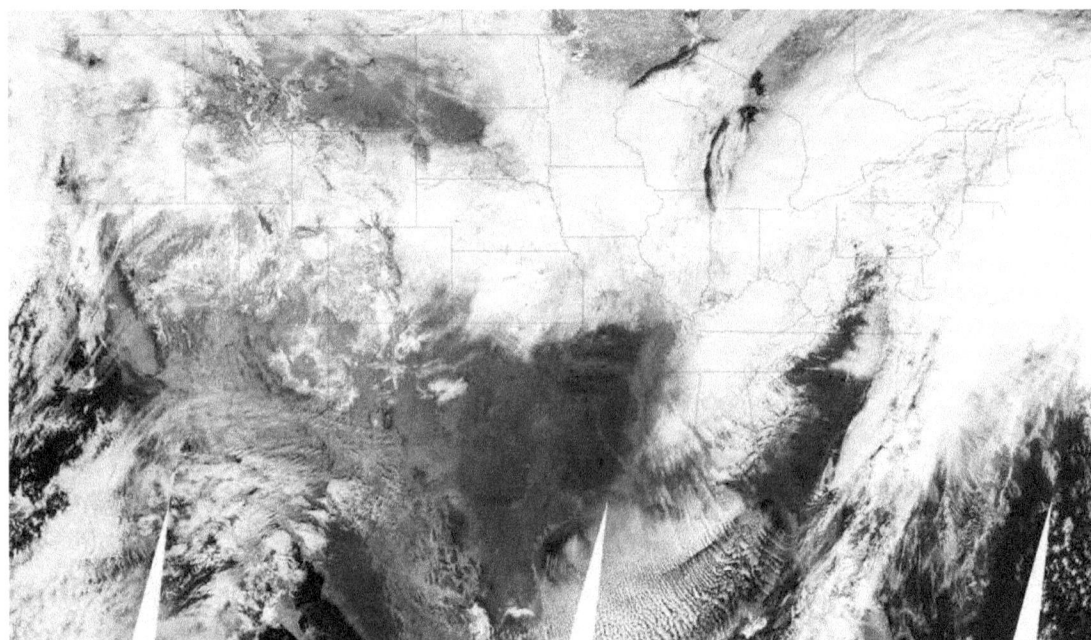

the significant advances in our understanding of climate change and climate variability made available through these publications, there would be no useful information to communicate to the public or to aid in decisionmaking.

In addition, CCSP produces an annual report to Congress that provides a yearly update on key scientific findings and plans for the coming fiscal year. CCSP also fosters stakeholder and community engagement by sponsoring workshops, such as the 2005 workshop on Decision Support that brought together experts and stakeholders for discussions of climate change and its impacts and yielded a publicly available report of its proceedings.[4] As mentioned above, CCSP also contributes substantial expertise and support to other national and international assessments, including the IPCC Fourth Assessment Report (IPCC, 2007a,b,c), the Arctic Climate Impact Assessment (ACIA, 2005), and many others.

Other key products of the program include the aforementioned 2003 CCSP Strategic Plan and a series of 21 synthesis and assessment products that are being produced during the 2006 to 2008 timeframe. The scope of the synthesis and assessment products was informed in part by the substantial stakeholder engagement in the earlier strategic planning process. These synthesis and assessment reports provide in-depth 'state of the science' information responsive to the CCSP overarching strategic

goals and related to specific national, regional, and sectoral issues (see Appendix 1 for a listing and description of these products, and <www.climatescience.gov/Library/sap/sap-summary.php> for information on available products and the status and timelines of products in preparation).

CCSP adds value to Federal agency efforts in climate change research and related activities by providing a structure and coordination mechanism for global change activities, resulting in strong leveraging of individual agency efforts through increased cooperation, collaboration, and the ability to jointly develop priorities that Federal agencies can consider in their planning. This facilitation and cooperation will continue and strengthen as CCSP improves its capability in communication and decision support, and will allow Federal global change endeavors to continue to do more than individual agencies could do separately, to increase the impact of accomplishments, and thus to provide additional worth to stakeholder communities and the public.

[4] See <www.climatescience.gov/workshop2005/finalreport>.

II | Climate Change Science Program Planning

II. Climate Change Science Program Planning

CCSP maintains an ongoing planning process to determine near-term objectives and interagency targets of opportunity as well as long-term strategic directions. This Revised Research Plan reflects the short-term (1-3 year) direction of the program, and thus provides an update for the period 2008 to 2010. It is an interim product of the ongoing CCSP strategic planning process.

To develop its near-term plans, CCSP utilizes a number of internal and external review processes to develop priorities, to assess progress, and to identify areas needing additional focus. CCSP principal representatives of participating agencies meet on a regular basis to discuss agency perspectives and priorities, to develop, implement, and evaluate the program's activities, and to assess progress. Individual member agencies develop and implement their own priorities that emerge from agency direction and agency stakeholder engagement, and they make a substantial contribution to the achievement of CCSP's strategic goals.

The program's interagency working groups (IWGs) also meet regularly to bring together the program managers responsible for implementing their individual agencies' research programs as they pertain to CCSP research, cross-cutting elements, and core approaches (a description of each IWG, including membership and

objectives, is available on the CCSP website <climatescience.gov>). Some IWGs convene Science Steering Groups composed of members of the relevant scientific communities as an additional means of increasing communication with those communities. IWGs also develop annual interagency implementation priorities in which they identify research activities that require cooperation and collaboration among member agencies and that are aligned with the science and cross-cutting elements defined in the Strategic Plan. These interagency implementation priorities are often multidisciplinary, highly integrative, and speak to not just one, but a combination of CCSP strategic goals. Implementation of these interagency priorities can result in joint funding solicitations, joint field campaigns, workshops, and other efforts. The combination of individual agency activities, interagency activities, international activities, and stakeholder engagement and outreach activities form the bulk of CCSP's progress toward its goals. Each year, the preparation of CCSP's annual report to Congress provides a means of self-evaluation as program accomplishments and gaps are reviewed relative to goals.

To inform its planning, CCSP also seeks, funds, and receives periodic evaluations from NRC *ad hoc* committees and standing committees established for that

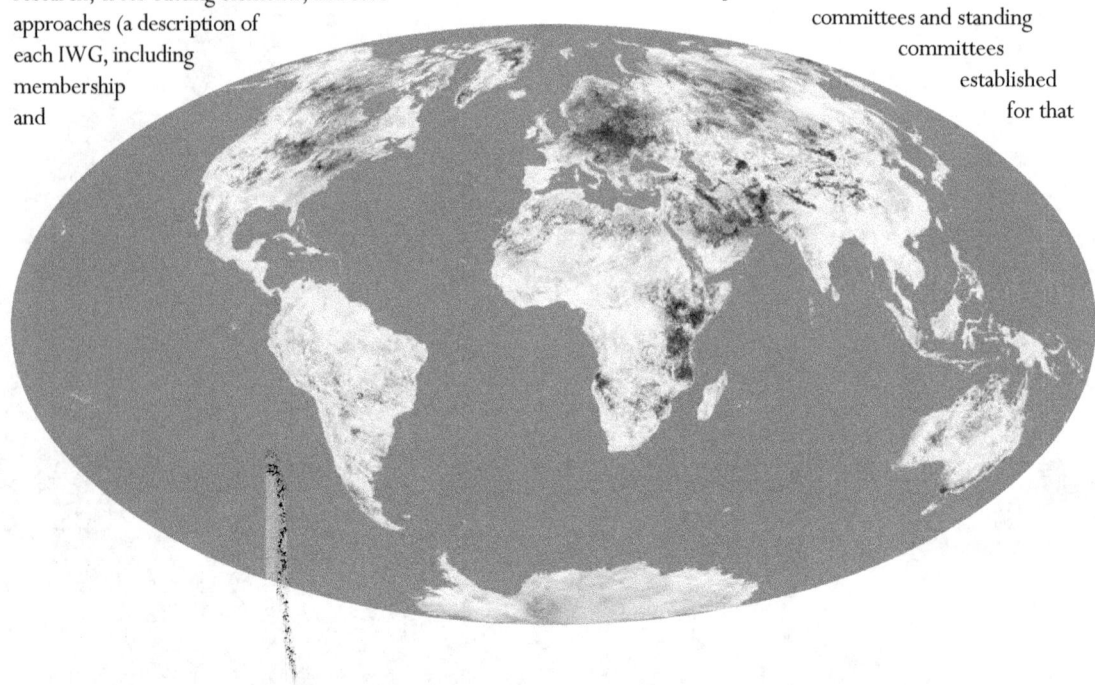

purpose. Recent NRC reports and assessments include *Thinking Strategically: The Appropriate Use of Metrics for the Climate Change Science Program* (NRC, 2005), *Evaluating Progress of the U.S. Climate Change Science Program: Methods and Preliminary Results* (NRC, 2007a), and *Analysis of Global Change Assessments: Lessons Learned* (NRC, 2007b) from the Committee on Strategic Advice on the U.S. Climate Change Science Program. CCSP also seeks stakeholder and community engagement and guidance through workshops, listening sessions, and public comment periods for key documents.

CCSP conducts long-term planning to determine the program's strategic directions. The last strategic planning effort resulted in the development of CCSP's 2003 Strategic Plan, which set the program's vision and goals for the period 2003 to 2013. In order to respond to the rapidly evolving needs of science and society, the program has now begun a new effort in long-range planning, with the goal of developing CCSP's next strategic plan.

The program has appointed an IWG charged with managing the strategic planning process. This group meets regularly to move toward the development of a new long-term strategic plan that will serve as the successor to the 2003 document. Among the tasks undertaken by this group are to assess progress, to identify areas of potentially greater emphasis for the future, to develop mechanisms for engaging a wide range of stakeholders and for informing planning through internal processes, and to begin to identify the building blocks that will comprise the new strategic plan. CCSP historically has utilized, and will continue to utilize, multiple inputs to inform its planning process:

- Stakeholder Input: CCSP will continue to utilize scientific advisory groups that support individual agencies, science steering groups organized to coordinate different CCSP research elements, and open dialog with the domestic and international scientific and user communities interested in global change issues. CCSP also conducts stakeholder workshops on specific topics [e.g., the 2005 Workshop on Decision Support (CCSP, 2005)]. As discussed in Section I, CCSP sees an immediate need for, and therefore intends to engage with, a broader range of more diverse consumers of CCSP's science than in the past, and to develop dialog and partnerships with a wide variety of stakeholders.

- The National Research Council: CCSP funds an NRC committee to provide independent, high-level, integrated advice on the strategy and evolution of the program. The committee's first report (NRC, 2007a) included findings and recommendations on the process for evaluating progress toward the five CCSP goals and a preliminary assessment of progress to date. The second report will identify priorities to guide the future evolution of the program in the context of established scientific and societal objectives. CCSP also relies on other NRC committees that focus on particular components of climate science and the climate system (e.g., NRC, 1999, 2000, 2001a,b; and the Climate Research Committee and the Committee on the Human Dimensions of Global Change) to help shape CCSP's programmatic evolution and prioritization. For example, the NRC recently produced a report on global change assessments at the request of CCSP (NRC, 2007b).

- **International Collaboration and Partnerships**: CCSP participates in and supports a wide range of international cooperative activities related to global change and climate change research, including WMO, UNEP, the World Climate Research Programme (WCRP), the International Geosphere-Biosphere Programme (IGBP), and the Global Earth Observations System of Systems (GEOSS), and relies on these activities to assist in providing global context to its efforts. By developing both conceptual and research frameworks, international research programs provide models that aid U.S. program managers in planning and coordinating their efforts, and help to leverage research and synthesis opportunities.

- **Climate Change Assessments**: CCSP supports and contributes to the IPCC assessment process as well as producing its own series of synthesis and assessment products. CCSP also participates in other assessments, including the development of a scientific assessment as required by the GCRA in parallel with this Revised Research Plan, and related activities that provide input to CCSP's on-going program development. The results of these syntheses help to define the program's future directions by discussing knowledge gained as well as identifying crucial knowledge gaps and assessments of impacts, risks, and opportunities.

At the highest conceptual level, CCSP's five strategic goals provide focus and facilitate programmatic integration across CCSP's 13 agencies and departments. These goals together with CCSP's four core approaches encapsulate the full range of climate-related issues, from basic observations, monitoring, and research to improve process understanding, modeling, and forecasting, vulnerability analyses, and other "human dimensions" of climate change, to decision support and communication of results and impacts. The program's detailed objectives, milestones, products, and payoffs, as articulated in the 2003 CCSP Strategic Plan, serve as foci for the implementation and integration of CCSP's overarching goals by its IWGs, which are organized around the program's identified research elements and cross-cutting elements. CCSP-participating agencies and departments coordinate their work through these discipline-related 'research elements,' which together support scientific observations and research across the gamut of interconnected issues of climate and global change. The four cross-cutting elements – observations and monitoring, decision support, communications, and international activities – address issues pertinent to all research elements.

III Highlights of
Progress: 2003 to 2007

III. Highlights of Progress: 2003 to 2007

Significant progress has been made in many areas of climate change research, as evidenced by the development of recent syntheses including the IPCC Fourth Assessment (IPCC, 2007a,b,c) and CCSP's synthesis and assessment products. Both of these efforts required the existence of relatively mature scientific research, substantial archives of environmental data, and robust modeling capabilities, a substantial portion of which were developed under CCSP auspices. Scientific progress has been substantial. The accomplishments of the past 4 years have led not just to advancement of scientific knowledge, but just as importantly, to the evolution and refinement of the science questions and approaches needed for current and future global change research, and to the improved delivery of robust scientific information for decisions and response strategies.

Substantive progress in CCSP Strategic Goals 1 through 3 is a required component of progress in many areas associated with Goals 4 and 5, since effective vulnerability analyses, human dimensions research, and decision support require reliable basic data, a clear understanding of processes, and projections whose uncertainties are well characterized. Key progress during 2003 to 2007 reflects this, with the greatest accomplishments showing in Goals 1, 2, and 3. In the coming years, CCSP will have both the need and the opportunity to advance research in Goals 4 and 5, while Goals 1 through 3 will continue to be important in order to provide a solid scientific foundation for breakthroughs in key areas of uncertainty and ever greater refinement of models, leading to better predictive capabilities and more robust tools for scientifically sound decision support, assessment of vulnerabilities, and management of risks.

What follows here is a brief, goal-by-goal summary of key accomplishments and progress by CCSP and its participating agencies. CCSP studies include those that are funded and reported by participating agencies either singly or jointly as part of their contribution to the program. More specific and detailed information on these accomplishments can be found in each issue of CCSP's annual report to Congress. The reader should note that the accomplishments discussed are examples and highlights only; they do not capture the full breadth and scope of CCSP research. They represent important progress on key questions, but in most cases they do not purport to be the definitive answers to these questions. Rather, these individual studies are part of the larger body of research that provides insights into important processes and that results in the evaluation, refinement, and evolution of both results and questions by the larger scientific community, which in turn moves science forward.

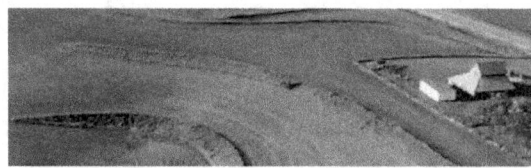

Goal 1: Improve knowledge of the Earth's past and present climate and environment, including its natural variability, and improve understanding of the causes of observed variability and change.

CCSP research requires a solid foundation of scientific observations, long-term monitoring data, and analyses of these data to provide a better understanding of Earth system processes, to understand Earth's past and present climate and the magnitude and extent of climate variability and change, and to test and to improve models. Analyses of collected observations and monitoring data underpin all aspects of climate system study. In the past few years, key analyses of collected data have provided important insights into understanding the nature and variability of the Earth system.

One example of a body of work with profound impact is the progress made in understanding temperature and moisture changes at continental scales, and in understanding the magnitude of climate change at high latitudes in the Arctic and Antarctic regions of Earth. Analysis of temperature and moisture records together with satellite images and ground-based measurements for North America and Europe show that both continents are experiencing earlier transitions from winter to summer, and that warmer spring temperatures are causing earlier spring green-up of vegetation and contributing to longer growing seasons overall (Dirmeyer and Brubaker, 2006). Observations in the western United States indicate that the annual peak in spring river runoff is occurring earlier in the season and is

supplying less water during the growing season (Mote et al., 2005). Satellite, airborne, and ground-based observations suggest that significant changes are occurring in the mass balance of the Greenland and Antarctic Ice Sheets that are inferred to be caused by warming at high latitudes (Rignot and Kanagaratnam, 2006; Velicogna and Wahr, 2006). New satellite-based observations of the polar regions indicate significant reductions in the volume of the Greenland Ice Sheet (Velicogna and Wahr, 2005), declining Arctic sea ice cover, and loss of ice mass in Antarctica despite no measurable change in snowfall over the last 50 years (Monaghan et al., 2006; Velicogna and Wahr, 2006). The IPCC Fourth Assessment concludes that it is likely that there has been significant anthropogenic warming over the past 50 years averaged over each continent except Antarctica (IPCC, 2007a).

Another key temperature finding comes from the body of analytical work reported in CCSP Synthesis and Assessment Product 1.1, *Temperature Trends in the Lower Atmosphere: Steps for Understanding and Reconciling Differences* (CCSP, 2006). Previously reported data showing discrepancies between the amounts of warming near the surface and higher in the atmosphere have been used to challenge the reliability of climate models and the reality of human-induced global warming. In these earlier observations, surface data showed substantial global-average warming, while early versions of satellite and radiosonde data showed little or no warming above the surface. Errors in the satellite and radiosonde data have been identified and corrected, resulting in better estimates of temperature trends. For recent decades, all current atmospheric data sets now show global-average warming that is similar to the surface warming. A next step is to use the current data and recent syntheses to improve climate-modeling results: while these observations are consistent with the results from climate models at the global scale, further work is needed to resolve remaining discrepancies in the tropics. Thus, while these observations and modeling results have provided increased scientific confidence in our understanding of observed climatic changes and their causes, areas of significant work remain.

It is critical to observe and understand variations in the average state of particular climate parameters like temperature and precipitation, but it is equally important to understand natural variability and changes in the frequency or intensity of extreme events like drought and unusually wet conditions. CCSP researchers have made progress in understanding the climate system's natural variability. One of the most well-known variability events

is the El Niño phenomenon, which recurs on a time scale of approximately 2 to 7 years and involves a warming of the eastern tropical Pacific in combination with changes in atmospheric circulation. Recent research links decadal changes in this pattern to droughts and wet conditions over North America and suggests that a portion of such decadal changes may be predictable (Seager et al., 2005). Additional recent accomplishments on the topic of drought used the geologic record to reconstruct drought history over the past millennium, using proxy records derived from tree rings and sediment cores (Cook et al., 2004). This research indicates that recent U.S. droughts in the west, while severe enough to cause substantial impacts to society, are relatively minor in comparison to naturally occurring droughts over the past 1,000 years. Interpreting changes in the characteristics of extreme events remains one of CCSP's ongoing research frontiers.

Additional work on atmospheric conditions provided an improved understanding of climate influences on ozone distribution. Using satellite measurements corroborated by surface measurements, a recent study found increases in ozone in the Antarctic middle stratosphere during Southern Hemisphere summer (December). Model simulations showed that these increases were caused by the delayed transition from dynamic springtime conditions to more stable summer conditions due to the springtime ozone hole. The lengthening of less-stable springtime dynamics forces the descent of ozone-rich air from higher levels of the atmosphere to the lower mid-stratosphere (about 30 km altitude). The same study also found that future greenhouse gas increases would produce similar ozone increases (Stolarski et al., 2006). Another study found that a doubling of atmospheric carbon dioxide (CO_2) caused a strengthening of the atmospheric circulation responsible for the global distribution of ozone. The results of this study indicate that total ozone will increase at high latitudes of the Northern and Southern Hemispheres, and decrease in the tropics (Jiang et al., 2007). Long-term data sets have been developed to provide information sufficient to test the skill of general circulation models at predicting radiative heating and cloud feedbacks (Mace, 2006a,b). The accuracy of measurements has been greatly improved to better evaluate model simulations (Turner, 2007; Turner et al., 2007).

Oceans, too, are experiencing change. Declines in Arctic sea ice have been observed both *in situ* and by satellites. Observations of global sea-level increases are consistent with the declining volume of land ice as well as observations of ocean warming, which contributes to sea-level rise by expanding ocean volume. Observations of the North Atlantic indicate a reduction in salinity (Curry and Mauritzen, 2005), which climate system models indicate may lead to a slowdown of the large-scale ocean circulation that transports heat to high-latitude regions (Stouffer et al., 2006a). Global-scale observations of ocean temperature indicate a pattern of warming that is generally consistent with climate model projections of greenhouse warming (Barnett et al., 2005). In the past decade, measurements from a variety of platforms, including satellites and ocean buoys, show warming in the top layers of the ocean, with strong evidence that the warming is due to increases in human-produced greenhouse gases (IPCC, 2007a). These data are significant because ocean heat storage is the largest component of the Earth's climate system for storing the energy imbalance between the sources and sinks of thermal energy. Even though the methods of observations are quite different, the fact that the magnitude and annual variability of the satellite-derived energy imbalance and the ocean heat storage match lends confidence to the interpretation of the underlying climate process.

The global ocean is a large and important carbon reservoir that regulates the uptake, storage, and release of CO_2, methane, and other climate-relevant chemical species to the atmosphere. The future biogeochemical behavior of this reservoir is quite uncertain because of the potential anthropogenic impacts on many ocean processes, in particular the impact of ocean circulation on carbon exchange and the impact of ocean acidification on the physiology, function, and structure of the complex and diverse ocean ecosystem. CCSP research has found that the ocean's ability to remove more CO_2 from the atmosphere will be impaired with warmer temperatures. Both the absorption of anthropogenic CO_2 and the deposition of acid rain from fossil fuel and agricultural emissions can contribute to the acidification of the global ocean, altering surface seawater acidity and inorganic carbon storage. Researchers have compared these inputs and concluded that (1) acid rain contributes a minor amount (2%) of acidity compared to the ocean uptake of anthropogenic CO_2, although this value likely represents an upper limit, and (2) the decrease in surface alkalinity from acid rain drives a net air-sea release of CO_2, reducing surface dissolved inorganic carbon (DIC). Total alkalinity and DIC changes mostly offset each other, resulting in a small increase in surface acidity. Additional impacts arise from atmospheric nitrogen deposition,

leading to elevated primary production and biological drawdown of DIC that in some places reverses the sign of the surface acidity and air-sea CO_2 exchange. On a global scale, the alterations in surface water chemistry from anthropogenic nitrogen and sulfur deposition are a few percent of the acidification, although the impacts are more substantial in coastal waters, where the ecosystem responses to ocean acidification could have the most severe implications for humans (Bates and Peters, 2007; Doney et al., 2007).

Goal 2: Improve quantification of the forces bringing about changes in the Earth's climate and related systems.

It is essential to understand the factors responsible for global environmental change in order to make long-term climate projections. These forcing factors include changing levels of greenhouse gases, land-cover changes, airborne dust and other particles (aerosols), and solar variability. CCSP research gives considerable attention to identifying and quantifying the effects of these forcing factors, and to understanding the ways in which factors cause feedbacks among Earth systems. As in the previous goal, the following examples of progress toward CCSP Goal 2 result from the integrated focus of multiple CCSP research elements.

Climate and the global carbon cycle are a tightly coupled system where changes in climate affect the transfer of atmospheric CO_2 to the terrestrial biosphere and the ocean, and *vice versa*. An important conclusion of recent carbon cycle research is that future warming is likely to lead to a further decrease in the efficiencies of land and ocean in absorbing excess CO_2, resulting in more of the emitted CO_2 remaining in the atmosphere (i.e., a positive feedback) (Fung et al., 2005). This assessment is based on advances in U.S. and global carbon observations and improvements in carbon cycle models. Controlled experiments on carbon uptake and release in ecosystems are one means of improving our understanding of carbon cycle dynamics, which can contribute to corresponding carbon cycle model improvements. For example,

Free-Air Carbon Dioxide Enrichment experiments, in which CO_2 is purposely injected into the air around a small plot of land, have led to the understanding that the mass of carbon in ecosystems initially tends to increase when exposed to increased levels of CO_2 (Jastrow et al., 2005; Norby et al., 2005). This increase may be limited by the availability of nutrients, although a comprehensive meta-analysis indicates that nitrogen supply generally keeps pace with plant demands in natural systems (Luo et al., 2006). Other controlled experiments in which ecosystems are purposely warmed generally indicate greater ecosystem CO_2 release with higher temperatures. However, there are still significant uncertainties associated with the biospheric response to climate change, particularly with respect to the complex and dynamic nature of ecosystems and their interactions with climate and the hydrologic cycle.

CCSP has also made significant advances in understanding the processes responsible for the production and destruction of other greenhouse gases, including methane and nitrous oxide. For example, recent analyses suggest that approximately 60% of all methane emissions from wetlands may occur in the tropics (Melack et al., 2004). In polar regions, recent studies have elucidated processes by which carbon, currently trapped either as organic matter or methane hydrates in the permafrost (frozen soil), is released to the atmosphere (Zimov et al., 2006). Warming increases these releases and can create an amplifying feedback loop of climate forcing. A new long-term, global reference for greenhouse gas stabilization scenarios and an evaluation of the process by which scenarios are developed and used was produced in 2007. This publication, Synthesis and Assessment Product 2.1, consists of two parts. Part A, *Scenarios of Greenhouse Gas Emissions and Atmospheric Concentrations*, uses computer-based scenarios to evaluate four alternative stabilization levels of greenhouse gases in the atmosphere and the implications for energy and the economy of achieving each level. Part A includes stabilization scenarios for six primary anthropogenic greenhouse gases – CO_2, nitrous oxide, methane, hydrofluorocarbons, perfluorocarbons, and sulfur hexafluoride – and uses updated economic and technological data and new tools for scenario development. Although these scenarios should not be considered definitive predictions of future events, they provide valuable insights for decisionmakers. Part B, *Global Change Scenarios: Their Development and Use*, examines how scenarios have been developed and used in global climate change applications, evaluates the effectiveness of current

scenarios, and recommends ways to make future scenarios more useful. Part B of the report concludes that scenarios can support decisionmaking by providing insights regarding key uncertainties, including future emissions and climate as well as other environmental and economic conditions.

With mounting concern that human activity is altering our environment, not just at the surface, but also in the extended atmosphere, it becomes increasingly important to improve our knowledge of all natural forcings on Earth. Over the past several centuries and longer term, the two main natural forcings have been solar variability (warming and cooling) and volcanic activity (cooling). The Solar Radiation and Climate Experiment (SORCE) mission has provided new insights on the solar forcing with SORCE making the most accurate measurements of the total solar irradiance (TSI) and the first daily observations of the solar spectral irradiance (SSI) in the visible and near infrared. During the past 11-year solar cycle, the Sun's increased brightness at solar maximum warmed the Earth's atmosphere by about 0.1°C in the 10 km nearest to the surface (the troposphere), 1°C near 50 km in altitude (at the top of the stratosphere), and 400°C at an altitude of 500 km in the thermosphere. Currently the Sun is at solar minimum, and the TSI observations indicate that the solar brightness has decreased by about 0.02% over the past decade (Woods and Lean, 2007). This result suggests that the solar forcing could reduce slightly the larger global warming trend due to human-induced greenhouse gases (e.g., burning of fossil fuels). Continued observations of the solar irradiance, such as by SORCE, are key to understanding the long-term solar changes that provide a natural forcing to climate change.

One of the largest uncertainties in projections of potential future climate change is the role of aerosols. Recent research has reduced some of this uncertainty, in part through efforts made possible by CCSP. In the first phase of preparing the synthesis and assessment report that deals with aerosol properties and their impacts on climate, a comprehensive paper has been published that reviews recent progress in characterizing aerosols and assessing their direct effect on climate change (Yu et al., 2006). This work, as well as other studies that will be published in 2008 as part of CCSP Synthesis and Assessment Product 2.3, *Aerosol Properties and Their Impacts on Climate*, served as a major resource in the preparation of the IPCC Fourth Assessment Report.

A diverse array of laboratory studies, field experiments, and modeling efforts have better defined aerosol formation processes, their properties, and their abundances. These studies involved multiple agencies, universities, observation platforms, and field teams deployed in multiple locations including Mexico City, the Gulf of Mexico, the Sahara, and other locations. Among the results of these activities are measurements that show a higher than expected formation of organic aerosols within the atmosphere, which could potentially have a cooling effect; demonstration of the influence of aging and composition on aerosol properties, and the ubiquity of absorbing (warming) aerosols and black carbon in the atmosphere; and the first well-sampled direct evaluations of the effects of a major Saharan dust storm on solar and thermal radiation across the atmosphere, which in turn has provided a way for researchers to test their understanding of how dust affects the radiant energy budget of the atmospheric column. This information is a key component in computer models that simulate both regional and global weather and climate. Taken together, the information resulting from these activities will enable more accurate calculation of aerosol influences on climate through their effects on absorption and scattering of light, as well as the indirect effect of aerosols through the alteration of clouds (both water and ice). This in turn will lead to more accurate model estimates of the climatic role of aerosols and a better understanding of radiative forcing of climate change by lower-atmosphere ozone and aerosols, which will ultimately lead to improved climate projections.

Land use and land cover affect the global climate system through biogeophysical, biogeochemical, and energy exchange processes. Variations in these processes due to land-use and land-cover change in turn affect local, regional, and global climate patterns. Key processes include uptake and release of greenhouse gases by the terrestrial biosphere through photosynthesis, respiration, and evapotranspiration; the release of aerosols and particulates from surface land-cover perturbations; variations in the exchange of sensible heat between the surface and atmosphere due to land-cover changes; variations in absorption and reflectance of radiation as land-cover changes affect surface reflectance; and surface roughness effects on atmospheric momentum that are land cover dependent. Human activity can and does alter many of these processes and attributes, but weather and climate, as well as geological and other natural processes, are also important. For example, land-cover changes such as deforestation and forest fires alter ecosystems and release CO_2, methane, carbon monoxide, and aerosols to the atmosphere. They also change the reflectivity of the land surface, which in turn determines how much of the Sun's energy is absorbed and thus available as heat, while vegetation transpiration and surface hydrology determine how this energy is partitioned into latent and sensible heat fluxes. At the same time, vegetation and urban structure determine surface roughness and thus air momentum and heat transport.

Examples of important research progress in understanding these factors and forcings include studies on the effects of urbanization and agricultural development on air temperature, and on the remote assessment of the effects of land-use variables linked to climate forcings in the midwestern United States and in Central America. For instance, researchers exploring the effects of urbanization and agriculture on regional climate found that irrigated agriculture in California tended to lower average and maximum near-surface air temperatures, while conversion of natural vegetation to urban areas tended to increase near-surface air temperatures. The surface temperature changes and their associated effects on the atmosphere also caused changes in the regional airflow. Overall, it was found that conversion of natural vegetation to irrigated agriculture has likely had a larger effect on the climate of California than urban growth, but increased conversion of irrigated land to urban/suburban development could alter this conclusion (Kueppers et al., 2007).

Similarly, crop type, yield, and land management affect the balance of greenhouse gas fluxes from land cover in the midwestern United States. The magnitude of surface changes such as tillage intensity affects residue cover and thus the moisture and radiation energy balances at the land surface through changes in evaporation and reflectance. However, these distinctions are difficult to assess across landscapes. Scientists working in Iowa have developed a method using Landsat Thematic Mapper and EO-1 Hyperion imaging spectrometer data to classify tillage intensity in cropland (Daughtry et al., 2006). A methodology has also been developed for observing changes in tropical forest cover over large areas using data with high temporal frequency from coarse-resolution satellite imagery. Proportional forest cover change is estimated from multi-spectral, multi-temporal Moderate-resolution Imaging Spectroradiometer (MODIS) data that are transformed to optimize the spectral detection of vegetation changes. This methodology has been applied using MODIS data in Central America. Landsat data are also used to record higher detail changes in forest cover in Central America. This work describes the distinct patterns of change from year to year due to land-cover changes resulting from forest clearing, regeneration, and changes in climate. It was found that the ability to detect forest cover change patterns using this methodology was relatively independent of the spatial resolution of the data. Associated model simulations indicated that the best metrics for detecting tropical forest clearing and regeneration are the shortwave infrared information from the MODIS data at 500-m resolution. Errors were found to range from 7 to 11% across the time periods of analysis (Hayes and Cohen, 2007).

Another important recent advance is the improved understanding of carbon cycling. A key component of the Nation's integrated carbon cycle research is the North American Carbon Program (NACP), the central objective of which is to measure and understand the sources and sinks of CO_2, methane, and carbon monoxide in North America and in adjacent ocean regions. Work conducted by NACP researchers and others has provided improved estimates of the amount of carbon being sequestered in North America and globally, and in particular, how the rate of carbon uptake is changing in all ecosystems. These estimates are made through the innovative combination of carbon cycle models and observations of carbon concentrations and isotopes (Fung et al., 2005).

An evaluation of North American carbon sources and sinks was generated as part of CCSP's Synthesis and Assessment Product 2.2, *The First State of the Carbon Cycle Report (SOCCR): North American Carbon Budget and Implications for the Global Carbon Cycle*. A key finding of this report is that, in 2003, North American terrestrial carbon sinks removed approximately 520 million metric tons of carbon per year (±50% with 95% confidence) from the atmosphere, which is equivalent to approximately 30% of North American fossil fuel emissions in 2003. Approximately 50% of the sink is due to the regrowth of forests in the United States on former agricultural land and on forested land recovering from harvest. This sink is expected to decline. As forests mature, they grow more slowly and take up less carbon from the atmosphere. The current source to sink imbalance of more than three to one (ratio of fossil fuel emissions to net terrestrial carbon uptake) and the potential trend of increasing sources and decreasing sinks suggest that addressing imbalances in the North American carbon budget will likely require actions focused on reducing fossil fuel emissions. Options focused on enhancing carbon sinks in soils and vegetation can contribute as well, but their potential is far from sufficient to offset current fossil fuel emissions. Work of this nature is important for assessing the efficacy of natural carbon uptake, as well as the potential for purposeful carbon capture in managed ecosystems (CCSP, 2007a).

Observations of ocean carbon are important for addressing uncertainties associated with the global carbon budget. New global scale ocean carbon analyses indicate increasing carbon concentrations in ocean water. In addition to confirming the oceans as a significant carbon sink, this information is also being used to estimate the increase in ocean acidity caused by increasing amounts of dissolved CO_2 and the potentially deleterious consequences for marine ecosystems (Orr et al., 2005). Recent measurements of carbon sedimentation along continental shelves have shown these regions to be responsible for a significant fraction of oceanic carbon uptake (Muller-Karger et al., 2005).

Recent climate warming has been particularly intense in boreal and Arctic regions, leading to concern that increasing air temperature in these ecosystems may indirectly increase the incidence of forest fires. Longer growing seasons lead to increased fire fuel loads, and increased temperatures can lead to drier conditions. In turn, fire provides rapid release of carbon and produces aerosols that affect atmospheric conditions. Beyond the emission of CO_2 and other greenhouse gases, understanding the consequences of large-scale fires for climate is challenging due to the many additional ways in which they influence the atmosphere and surface. A recent study in Alaska found that there was intensification in the climate warming in the first year after a major fire but a slight decrease in the local climate warming when averaged over the 80 years of the study. The long-term result, which was primarily due to plant regrowth increasing the summer reflectivity of the burned surface, appeared to be more significant than the fire-emitted greenhouse gases (Randerson et al., 2006). The study results suggest that possible future increases in wildfire in some parts of the boreal zone of Alaska may have different feedbacks to global warming than previously thought.

Scientists are also studying the possibility that increased permafrost thawing due to warming in Arctic regions could cause the release of substantial amounts of carbon long held in the frozen tundra. There appear to be two potential mechanisms for the carbon to reach the atmosphere: drainage of the carbon-rich river flow into the Arctic Ocean with subsequent emission, and direct respiration or recycling of the newly released carbon. Measurements made in the Yukon River Basin in northern Canada have shown that the latter process predominates (Guo and Macdonald, 2006). These results have significance for understanding both the movement of carbon through Arctic landscapes, and the potential effects of that carbon on ecosystems and the atmosphere.

CCSP's interdisciplinary research on the carbon cycle has also produced a set of analyses using long-term observations of several young and mature forests. Results from this work show that forest carbon storage has been increasing in these ecosystems and is not in balance with the carbon lost by respiration and decay. This result is contrary to the contemporary concept of near balance of carbon sources and sinks in mature forests (Zhou et al., 2006). The gain in forest carbon is typical of findings from U.S.-based large-scale networks, as well as observations made in mature forests in China. Evidence is therefore mounting that these sinks for atmospheric CO_2 offer significant potential for modulating the rate of atmospheric CO_2 increase (Urbanski et al., 2007).

Goal 3: Reduce uncertainty in projections of how the Earth's climate and related systems may change in the future.

Uncertainty provides a measure of the confidence that can be placed in forecasts of future climate and related Earth systems. Reducing uncertainty is crucial to providing decisionmakers with useful, reliable tools for assessing strategies for adaptation, mitigation, and other forms of risk reduction. However, the wording of the goal is in-complete: "reducing uncertainty" is only part of the story. Improving the projections themselves and understanding both the nature and implications of uncertainties are equally important, as noted in the NRC Metrics report (NRC, 2005). The thrust of this goal is therefore to improve projections, and to characterize their uncertainty, in order to improve the utility of projections of how the Earth's climate and related systems may change in the future in response to natural and human-induced forcings. CCSP has significantly advanced the ability to estimate future Earth system conditions at time scales ranging from months to centuries and at spatial scales ranging from regional to global. The primary tools for Earth system prediction and projection are computer models that reflect the best available knowledge of Earth system processes. Reducing uncertainty requires continual

integration of observations and modeling across the full range of climate and Earth systems research.

When recent model simulations of the climate of the past 100 years are compared to observations, the results generally indicate improvements over previous generations of models, including the ability to represent weather systems, climate variability (e.g., monsoons, El Niño), ocean processes (e.g., the Gulf Stream), surface hydrology, and other Earth system processes, components, and dynamics (Collins et al., 2006; Schmidt et al., 2006). One of the ways in which these models have advanced is through improvements in the representation of the processes responsible for key Earth system feedbacks such as those associated with water vapor, clouds, sea ice, and the carbon cycle (Delworth et al., 2006; Gnanadesikan et al., 2006; Wittenberg et al., 2006). A significant improvement in representing the vertical cloud structure in mixed-phase clouds in climate models was achieved by explicitly treating ice formation and liquid conversion to ice via the Bergeron-Findeisen process (Liu et al., 2007).

For a model to produce a realistic climate projection, it must include realistic representations of physical parameters such as cloudiness, precipitation, and solar spectral irradiance. Understanding the influence and feedbacks of clouds on climate has proven to be a challenging task for scientists worldwide, but progress is being made. Recent innovative studies using newly developed, detailed models of cloud processes coupled with a global climate model provide results that are significantly more consistent with observations than traditional cloud-modeling techniques (Ovtchinnikov et al., 2006). The incorporation of improved cloud representation in climate models is expected to reduce the uncertainty in predictions of the global and regional water cycle and surface climate.

Clouds are a major component in the global reflectance of sunlight, and energy from the sun not reflected back to space provides the driving energy to Earth's weather and climate systems. Year-to-year variability in the global reflectance is dominated by the variability of cloudiness in the tropics; but, on the other hand, scientists have recently found little change in the year-to-year variability of reflectance at middle and high latitudes despite decreases in the highly reflective snow and sea ice cover. This result appears to be due to the compensating increase in cloud cover balancing the decreasing surface-level reflectance (Loeb et al., 2006). Clouds continue to provide the largest source of uncertainty in model estimates of climate sensitivity, although a recent study finds evidence that, in most climate models used in the IPCC Fourth Assessment Report, clouds provide a positive feedback (Soden and Held, 2006).

The magnitude of future warming will be strongly influenced by both the extent to which atmospheric water vapor concentration increases in response to an initial warming caused by increases in CO_2 and other greenhouse gases and the extent to which increasing water vapor affects clouds and their radiative properties. An accurate representation of this feedback in climate models is critical for making long-term climate projections. Recent innovative analyses have shown that water vapor increases in the upper atmosphere measured by satellites and balloon-borne sensors are generally consistent with state-of-the-art climate model simulations, lending credence to the ability of current models to represent the water vapor feedback (Fu and Johanssen, 2005).

Analyses of climate model simulations generated for the IPCC Fourth Assessment have identified several additional characteristics of climate change projections common to all of the models (Held and Soden, 2006). Examples of these robust model projections include strong subtropical drying, weakening of large-scale tropical atmospheric motions, and expansion of the poleward upper atmospheric wind pattern known as the Hadley circulation. In another study, several models were used to investigate the effects of the freshwater input from melting ice and glaciers on the currents in the North Atlantic (Stouffer et al., 2006a). These currents are important due to their large-scale transport of heat. The study concluded that, in response to expected levels of freshwater input in the northern North Atlantic, the average modeled large-scale deep ocean current weakens by about 30% by the end of the century. All models simulate some weakening of this deep circulation, but no model simulates a complete shutdown.

CCSP researchers also use the geological record to test and apply climate models, particularly in cases where that knowledge has a bearing on climate change processes relevant to current society. For example, the mid-Pliocene provides an unequaled paleoclimatic laboratory to test the sensitivity of the physical models that we rely upon for estimating the magnitude and spatial variability of future warming. Estimates of mid-Pliocene (~3.0 million years ago) global warming suggest that temperatures were 2°C greater than today (Haywood and Valdes, 2004; Dowsett et al., 2005). This level of warming is within the range of the IPCC estimates of global temperature increases for the 21st century, and no other time period in the past

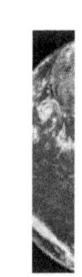

3 million years approaches this level of warming. Scientists have identified many of the primary forcing mechanisms that contribute to the current global warming, but there is uncertainty about the relative impact of each forcing and associated feedbacks. The mid-Pliocene presents the reverse situation: global data sets reveal the mature state of a warmer world, but the forcings that led to Pliocene warming are only partially identified. Analysis of data sets compiled by the Pliocene Research, Interpretation and Synoptic Mapping group (PRISM) suggests that a combination of increased greenhouse gases and increased ocean heat transports acted concurrently through undetermined feedback relationships (Dowsett et al., 2005; Haywood et al., 2005, Chandler et al., 2008). To fill in the picture, data are just beginning to emerge that describe the deep ocean state during the Pliocene (Cronin et al., 2005). Understanding the climate distribution and forcing for the Pliocene period may help improve predictions of the likely response to increased atmospheric CO_2 in the future, including the ultimate role of the ocean circulation in a globally warmer world.

The CCSP modeling strategy utilizes a multi-tiered approach in which new and improved Earth system sub-models (e.g., clouds, ecosystem dynamics, sea ice) are developed and tested by individual researchers or small research teams. When significant improvements in these sub-models arise, they are integrated as appropriate into high-end Earth system models. A result of these ongoing efforts is a set of U.S. models that expand beyond earlier atmosphere-ocean models to include relatively sophisticated representations of land-surface hydrology, sea ice, ecosystems, and atmospheric chemistry. U.S. Earth system modeling centers have used variations of these models to produce ensembles of projections that are providing important new perspectives on potential future climate system change (Meehl et al., 2004a; Stouffer et al., 2006b). These ensembles are also being used to characterize the intrinsic uncertainty associated with potential future climate change.

Because Earth system models are extremely complex and benefit greatly from input and evaluation by multiple research teams, several new efforts have been initiated to enable sharing, testing, and improvement of these models by diverse groups of researchers (Meehl et al., 2004b, 2005). Many of the recent model simulations referred to above are now widely available through a new capability for data archiving and dissemination developed by the Program for Climate Model Data and Intercomparison (see <www-pcmdi.llnl.gov>). Large strides have been made in creating climate model code according to a set of standards that facilitate exchange of sub-models (e.g., the Earth System Model Framework, <www.esmf.ucar.edu>), which enables researchers to readily trace the source of differences between various models and between models and observations.

The projections made by CCSP research pertain not just to physical climate, but also to other components of the Earth system, including atmospheric chemistry. Continuing research has provided an estimate that the recovery of the Antarctic ozone hole will occur approximately 10 to 20 years later than the previous estimate of 2050 (Newman et al., 2006). As a result of the Montreal Protocol and its amendments, the use of ozone-depleting substances (ODS) has been greatly reduced. Improved understanding of atmospheric dynamics now gives 2001 as a better estimate of when the ODS peak occurred in the Antarctic stratosphere. This date is later than had been estimated previously and results in a longer projected time scale for recovery back to pre-1980 (unperturbed) levels of ODS.

Goal 4: Understand the sensitivity and adaptability of different natural and managed ecosystems and human systems to climate and related global changes.

Goal 4 is an area of need and opportunity for CCSP over the coming years, in keeping with the increased availability of results from Goals 1 through 3 as inputs. Significant advances have been made in understanding the potential impacts of climate change, and in the improvement of methodologies as new information becomes available. CCSP research typically uses many different sources of information, including analyses of paleoclimate data, direct monitoring and observations, process studies, and model-based projections. Increasingly, research also accounts for the dynamic nature of the response of human and natural systems to climate change.

Research to understand ecosystem and human system sensitivity and adaptability encompasses and integrates a

wide range of potential impacts on societal needs such as water, human health, and agriculture, as well as potential impacts on natural terrestrial and marine ecosystems. This integration is exemplified by the development of several of CCSP's synthesis and assessment products (SAPs; see Appendix 1 for SAP titles and brief descriptions). These products address a range of issues including sensitivity of coastal areas to sea-level rise; effects and impacts of climate change on agriculture, energy, transportation, and human health and welfare; and thresholds of change and adaptation options for ecosystems and resource sectors.

The potential implications of sea-level rise point to the need to account for a wide variety of factors when assessing future impacts. For example, some measures to protect coastlines may have negative side effects, such as the potential for wetlands loss when inland barriers are constructed, preventing the wetlands from migrating inland in response to rising sea level (Cahoon et al., 2006). Similarly, the adaptability of complex coastal ecosystems is becoming better understood. For example, studies in a Chesapeake Bay marsh ecosystem showed that rising sea level, increasing atmospheric CO_2, and high rainfall can interact and improve the growth of a relatively tall bulrush at the expense of a hay-like cordgrass that grows in thick mats (Erickson et al., 2007). The thick cordgrass mats trap sediment and organic material more effectively than do the bulrushes, and so bulrush-dominated marshes are less able to rise in elevation through the addition of sediments. Such changes in species composition, caused by interacting global change factors, can therefore influence the ability of coastal marshes to adapt and keep abreast of sea-level rise.

Carbon cycle scientific research and modeling are highly relevant to carbon management needs, as demonstrated by a recent study that estimated the spatial variability of net primary production and potential biomass accumulation over the conterminous United States (Potter et al., 2007). This study's model-based predictions indicate a potential to remove carbon from the atmosphere at a rate of 0.3 GtC yr[1] through afforestation of low-production crop and rangeland areas. This rate of carbon sequestration could offset about one-fifth of the annual fossil fuel emissions of carbon in the United States. Another study (SAP 2.2; CCSP, 2007a) found that terrestrial ecosystems of North America are assimilating about 30% of the CO_2 emissions from fossil fuel sources.

Observational and modeling studies of terrestrial ecosystems indicate a wide variety of changes in which it appears that climate variations play a significant role. For example, recent evidence indicates a northward expansion of the ranges of many bird and butterfly species in the United States corresponding to warming in the region (Sekercioglu et al., 2004). Satellite and *in situ* observations also indicate a trend toward earlier growth of spring vegetation (Angert et al., 2005). In addition to temperature and hydrologic changes, the increasing level of atmospheric CO_2 is now thought to play a role in changing ecosystem distributions and characteristics due to its fertilizing effect. Agricultural yield models account for this effect, and project a range of agricultural impacts depending on the magnitude and nature of future climate change, crop types, and the types of adaptive measures that are adopted. Recent research indicates that different strategies than those currently in use may be required to manage insects, weeds, and diseases in agricultural systems (Ziska and Runion, 2006).

Another example of the ecological consequences of climate change involving insects and affecting adaptability is the devastation of millions of hectares of western U.S. and Canadian pines by bark beetles during the warmth and drought of 2000 to 2004. Recent modeling and observations reveal that beetles invading northernmost lodgepole pine trees are now only a few kilometers from previously pristine jack pine populations (Logan and Powell, 2007). This may create a pathway for bark beetles to reach valued pine forests in the eastern United States and Canada.

CCSP's integrated approach to understanding the sensitivity and adaptability of natural ecosystems to climate change has also been applied in remote, high-latitude regions in both hemispheres. In the Southern Hemisphere, the West Antarctic Peninsula is experiencing some of the most rapid warming on Earth, which is causing loss of sea ice and increased snow precipitation. In turn, these changes are having major contrasting impacts on the adaptability of different penguin species. For example, in the vicinity of Anvers Island near the West Antarctic Peninsula during the last 3 decades, populations shifted south, so that local abundance of the ice-dependent and snow-intolerant Adelie penguins decreased by 65% (currently about 5,000), while the abundance of Chinstraps and Gentoos increased by 2,730 and 4,600% (currently about 300 and 650), respectively (Ducklow et al., 2007). In the Northern Hemisphere, climate warming has caused significant declines in total cover and thickness of sea ice and progressively earlier ice breakup in some areas. These changes have been linked to increasing vulnerability of polar bear populations, causing them to extend their normal fast for longer periods during the open-water season (ACIA, 2005; Stirling and Parkinson, 2006).

Components of CCSP research funded in part through the U.S. Joint Global Ocean Flux Study program have explored ecosystem impacts in the open ocean resulting from climate variability and change as well as from changes in ocean chemistry and thermal structure. An example of a chemical impact is the chain of events causing the oceans to become more acidic due to the absorption of increasing concentrations of atmospheric CO_2 (Orr et al., 2005). Ocean warming tends to increase vertical stratification (layering) and thus slow the overturning of nutrient-rich deep-ocean waters (Schmittner, 2005). Recent model projections suggest that increased ocean acidification and increased layering of the upper ocean due to warming are likely to reduce plankton production. These model results are supported by satellite observations indicating significant changes in photosynthetic plankton concentrations, including declines in the North Atlantic and Pacific and increases in the Indian Ocean (Gregg et al., 2003).

In addition to managed ecosystems, CCSP research has expanded understanding of the sensitivity and adaptability of a variety of other sectors of societal interest. One of these is human health, which may be affected directly by changes in temperature and storm intensity, or indirectly through changes in distributions of insects or other vectors that carry pathogens. An example of research in this area is the effects of climate change on heat waves. Recent observational and modeling work suggests that the probability of heat waves such as the one that occurred in Europe in 2003 has increased significantly, and that future warming may make heat waves of similar magnitude a normal summer occurrence within several decades (Meehl and Tebaldi, 2004). Recent research on the societal dimensions of climate variations has shown that, to be valuable for human health decisionmaking and other societal areas of interest, physical climate analyses such as the aforementioned study of heat waves must be assessed within a complex fabric of other social and environmental factors (Poumadere et al., 2005). An example is the general increase in financial losses due to hurricanes over the past century, which may be attributable more to expanding coastal development than to changes in hurricane characteristics (Pielke et al., 2005). In regions such as central Africa, where the capacity to adapt to environmental variations is often relatively low, recent research has shown strong correlations between year-to-year climate variations and malaria outbreaks (Thomson et al., 2006).

These are a few examples of CCSP's research examining the sensitivity and adaptability of human and natural systems to climate variability and change. It is clear from this work that climate variations can have both beneficial and adverse effects on environmental and socioeconomic systems. However, future projections indicate that the larger the magnitude and rate of climate change, the more likely it is that adverse effects will dominate (NRC, 2002).

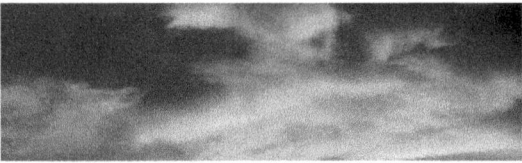

Goal 5: Explore the uses and identify the limits of evolving knowledge to manage risks and opportunities related to climate variability and change.

New knowledge emerging from the Nation's basic research on global environmental variability and change has created a number of important opportunities for applying that knowledge to help decisionmakers make good choices supported by sound science. Over the past several years, CCSP has taken three main approaches to explore and communicate the potential uses and limits of this knowledge: 1) the development of scientific syntheses and assessments; 2) supporting and exploring adaptive management and planning capabilities; and 3) development of methods to support climate change policy inquiries. Some key examples of progress are provided below.

The key focus of the program's assessment activities is its current suite of 21 synthesis and assessment products, which are intended to provide current evaluations of the science foundation for use in informing public debate, policy, and operational decisions, and for defining and setting the future direction and priorities of the program. These reports examine various aspects of climate change science and impacts on a national, regional, or sectoral basis as appropriate to each topic. A listing of synthesis and assessment products is provided in Appendix 1, and a description of each product and its schedule for completion can be found at <www.climatescience.gov/ Library/sap/sap-summary.php>.

Another important focus for the program's synthesis and assessment activities has been its involvement in the IPCC and other international assessment activities, including providing coordination and support to the IPCC Fourth Assessment Report. IPCC's major activity is to prepare at regular intervals comprehensive assessments of policy-relevant scientific, technical, and socioeconomic information appropriate to the understanding of human-induced climate change, potential impacts of climate change, and options for mitigation and adaptation. For the IPCC Fourth Assessment (2007), approximately 120 U.S. scientists were IPCC authors and 15 Review Editors. The United States co-chaired and hosted IPCC Working Group I, which primarily addressed physical science aspects of climate change. The United States has also played significant roles in the WMO/UNEP ozone assessments (e.g., WMO, 2003), the Arctic Climate Impact Assessment (ACIA, 2005), and the Millennium Ecosystem Assessment (MEA, 2005), among others.

The second of CCSP's decision support approaches is the exploration of adaptive management strategies. Activities under this approach develop and evaluate options for adjusting to variability and change in climate and other conditions through 'learning by doing' and integrating knowledge with practice. This area of work grows out of the insight that a key to assessment and decision support is close and ongoing interaction between users and producers of information. While this area is less well developed than some others in CCSP's portfolio, it is a strong and integral component of the missions of several CCSP member agencies, and those agencies provide extensive stakeholder engagement and decision support for adaptation, mitigation, and management of risk to policymakers and land managers at national, regional, state, and local levels.

As one example of the development of methods to support climate change policy inquiries, CCSP scientists developed and documented a 'water supply stress index' that calculates water shortage risks across the conterminous United States. The index is based on models and observations that integrate the effects of climate, land cover, and current water uses by municipalities and industries on water supply (Sun et al., 2006). The water supply stress index and the methods associated with it will be used by local and regional decisionmakers to quantify the likelihood of future water shortages under changing climate, water, and land uses in order to determine adaptation practices. Incorporation of the subsurface water table into regional climate models is important, since land-cover changes

produce significant effects on the water table and the hydrologic cycle. Shallow water tables can be either a sink or source of water to the surface soil depending on the relative balance of infiltration versus evaporation (Fan et al., 2007). Recent studies using detailed observations and regional climate models have confirmed the long-held anecdotal understanding that the fraction of rainfall that either recharges groundwater or ends up as streamflow tends to decrease when the fraction of land devoted to agriculture increases. This result suggests that intensive agriculture can amplify surface water stresses, particularly during drought conditions (Jayawickreme and Hyndman, 2007).

Another example of this work is an ongoing project that brings together researchers who study climate processes and their effects on the U.S. Southwest with individuals and organizations that need climate information to make informed decisions (Jacobs et al., 2005). Numerous tangible benefits from this project have helped a wide variety of decisionmakers, from state and local water planners to farmers to public health officials. For example, the project developed a suite of products that make predictions of water availability months in advance, allowing water managers to adjust reservoir levels accordingly to meet the competing demands for this scarce resource.

Yet another example is the combined use of satellite-based observations of fires and moisture conditions together with seasonal climate forecasts to provide information to fire managers to help them make early and effective decisions about the resources they will need to cope with emerging fires and fire-season dangers. One way in which this information is communicated is through annual workshops targeted separately at eastern and western U.S. fire hazards, which bring together climate scientists and forecasters with fire managers to produce seasonal fire outlooks (see <www.ispe.arizona.edu/climas/conferences/NSAW>). There are many other examples of the exploratory use of seasonal-to-interannual climate information for decisionmaking both domestically and internationally.

In addition to CCSP's work on adaptive management at seasonal-to-interannual time scales, the program is also developing valuable information for long-term (decades to centuries) adaptation issues. One example is the program's analyses of ways in which agricultural practices might be adjusted to take advantage of rising CO_2 levels and to cope with potentially warmer temperatures and decreased moisture availability. Recent work has shown that sufficient variability exists within some crop species

to begin selecting for crop varieties that could maintain or increase yields in a future enhanced-CO_2 environment (Boote et al., 2005).

An example of regional decision support is the work carried out by the Consortium for Atlantic Regional Assessment (CARA), which is providing data and tools to help decisionmakers understand how outcomes of their decisions could be affected by potential changes in both climate and land use. On an interactive, user-friendly web site, CARA has organized data on climate (historical records and future projections from seven global climate models), land cover, and socioeconomic and environmental variables to help inform local and regional decisionmakers (see <www.cara.psu.edu>). The CARA tools and tutorials are designed to help decisionmakers understand the issues related to land use and climate change by gathering, organizing, and presenting information for evaluating alternative mitigation strategies.

A workshop involving scientists and managers, co-led by several CCSP agencies under the auspices of the U.S. Coral Reef Task Force, resulted in the publication of *A Reef Manager's Guide to Coral Bleaching* (Marshall and Schuttenberg, 2006). The combined results of research on coral bleaching among state/territorial, Federal, academic, nongovernmental, and international scientists concluded that warming sea surface temperatures are a key factor in mass coral bleaching events. The guide provides managers with strategies to support the natural resilience of coral reefs in the face of climatic change.

CCSP's third decision support approach is to help inform inquiries related to climate change policy, in part by using comparative analyses of climate change scenarios. Several CCSP-participating agencies work extensively with scenarios and scenario development (for further information on these activities please see specific agency narratives in *Our Changing Planet*, CCSP's annual report to Congress). One example is a collaborative effort between climate scientists and New York City water infrastructure planners that is using regional scale hydrologic scenarios to inform the long-lasting investments that are being considered in the modernization of the city's water supply system (see <www.ccsr.columbia.edu/cig/taskforce>). Another example is the application of carbon cycle research to assess the potential feasibility, magnitude, and permanence of a variety of different carbon sequestration options, as noted in the U.S. Carbon Cycle Science Plan (Sarmiento et al., 1999). An initial result from this line of work is the preliminary conclusion that the restoration of inland wetlands could be a particularly efficient component of carbon sequestration in North American prairie lands (Euliss et al., 2006).

CCSP researchers have also developed new metrics for estimating greenhouse gas emissions and carbon sequestration in the agricultural and forestry sectors (Birdsey, 2006). These sectors can reduce atmospheric greenhouse gas concentrations by increasing carbon sequestration in biomass and soils, by reducing fossil fuel emissions through use of biomass fuels, and by substituting agricultural and forestry products that require less energy than other materials to produce. The new metrics are being used as the basis for reporting greenhouse gas information from the agricultural and forestry sectors (see <www.eia.doe.gov/oiaf/1605/frntvrgg.html>).

In the area of climate change and human health, a team of climate and health scientists funded by CCSP-participating agencies has been conducting interdisciplinary research on climate variability, climate change, land use, air quality, and human health in the New York Metropolitan Region. They have developed an integrated modeling system for assessing potential public health impacts related to heat stress and ground-level ozone. In particular, they linked a global climate model to regional climate models and a regional atmospheric chemistry model to produce downscaled heat and ozone estimates for the 1990s and the 2020s, 2050s, and 2080s under fast- and slower-growth socioeconomic and climate scenarios and down-scaled land-use scenarios. A health risk assessment of mortality impacts was conducted at the county level with these multi-model simulations of future environmental conditions. As part of this effort, more recent research has focused on providing improved understanding of climate and health vulnerability for stakeholders in support of decisionmaking in the New York region. This in-depth climate-health research is providing improved tools for decisionmakers in the region regarding public health risks due to potential heat and air quality changes (Kinney et al., 2006).

CCSP scientists have also been working with the Centers for Disease Control and Prevention and partners in environmental public health to provide environmental data products that would be of benefit to the Environmental Public Health Tracking Network (EPHTN). EPHTN will establish a national network of local, state, and Federal public health agencies to track trends in priority non-infectious health effects. This effort is being undertaken as part of the Health and Environment Linked for Information Exchange in Atlanta project, and is demonstrating a process for developing a local environmental public health tracking network. In a 2006 report, it was found that augmenting the U.S. Environmental Protection Agency Air Quality System observations with National Aeronautics and Space Administration MODIS-derived PM2.5 (particulate matter that is 2.5 mm or smaller in size) observations increases the temporal and spatial resolution of fine particulate estimates. The report also found that such augmentation also increases the accuracy in estimating concentrations of an environmental hazard such as PM2.5, which is important for environmental public health tracking. High concentrations of PM2.5 are associated with adverse health reactions (e.g., respiratory and cardiovascular problems) (Rosen et al., 2006).

Another important way in which CCSP is helping to inform climate change policy inquiries is through integrated assessment modeling, which considers the social and economic factors that may lead to climate change (e.g., greenhouse gas emissions) and the resultant effects of those activities on the Earth system and human welfare. These models are useful for considering the costs and effects of various policy options. One important result of this work suggests that reducing emissions of greenhouse gases other than CO_2 could be an economically efficient first step in reducing the overall atmospheric burden of greenhouse gases (Hansen and Sato, 2004).

These examples collectively provide a sketch of the progress CCSP has made toward its goals; additional information on these items and others can be found in CCSP's annual report to Congress. While these accomplishments, taken together, represent significant progress in climate change science and constitute a substantial portfolio of work, many questions remain to be answered within each of CCSP's strategic goals. In addition, the pertinent research questions have evolved in keeping with information gathered and knowledge gained, and the changing needs of society. The remainder of this Revised Research Plan provides a discussion of these questions and the plans that CCSP envisions as a means to answer them.

IV

Emerging Priorities for the Climate Change Science Program

IV. Emerging Priorities for the Climate Change Science Program

Any scientific research program must evolve over time based on what has been learned during earlier periods, and CCSP is no exception. This is particularly true for an Earth science-related program, in which the past several years have brought dramatic increases in knowledge; significant advances in the length and quality of observational data sets (including more comprehensive observations of climatic phenomena than were previously possible); improvements in the scope, resolution, and quality of models; significant advancements in computing capabilities for use in implementing more complex and higher resolution climate and Earth system models; and the initiation of several major observational efforts that have only now begun to yield results for integrated scientific study, or will appear shortly in the near future.

One of the most significant advancements of recent years is that ongoing monitoring of key Earth systems and analysis of records extending back through time (including paleoclimate information gleaned from studies of geologic materials and other climate proxies) have revealed evidence for a number of important Earth system changes and previously unknown processes, including (but not limited to) such things as:

- The continuation of the trend toward higher global average surface temperature manifested both on land and in the oceans
- Changes in atmospheric composition that both contribute to climate change and that are affected by climate change
- Changes in the cryosphere (e.g., Arctic sea ice coverage, significant changes in ice mass in Greenland and Antarctica, and permafrost temperature)
- Sea-level rise
- Changes in patterns and frequency of wildfire
- Changes in species distributions
- Ocean acidification and its consequences
- Changes in storminess

- Hydrologic changes affecting both humans and ecosystems, including changes in water availability and quality, glacier retreat, timing of snow melt, and potential for flooding events and/or drought

- Unexpected variations in seasonal greenness in tropical and temperate forests

- Variability of climate and the potential for abrupt change.

For many of these changes, attribution to causative factor(s) is not clearly established, and accelerated research and analysis are needed to identify more clearly the scientific bases that underpin such changes, including a means for testing hypotheses of potential attribution.

In the 4 years since the 2003 Strategic Plan was published, the climate community has also completed work on several important assessments, including the IPCC Fourth Assessment Report (to which CCSP made substantial scientific contributions) and the synthesis and assessment products being developed under the auspices of CCSP and discussed in the preceding section. These assessments have helped to integrate related scientific areas, to provide a comprehensive report on the state of the science, to address the causes and consequences of climate change, and to examine their effects on risk and vulnerability of Earth systems and human communities. These assessments are intended to provide input to the broader climate policy community, and to shape external dialogues and to frame the new questions that face land

and resource managers, communities, and policymakers. Discussions within the user community have highlighted the need for CCSP to provide more regionally resolved and sector-specific information about climate through downscaling of models, tools, and research results (including societal impacts of and vulnerabilities to climate change), and to provide the rigorous scientific basis to support increased societal planning for adaptation to and mitigation of the effects of climate change.

Taken together, the factors discussed above have led to a need for evolution in both research questions and approaches, and they have exerted a significant influence on CCSP's research directions. The past several years have seen dramatic increases in knowledge about climate and Earth systems; significant advances in both the length and quality of observational data sets from ground-based, airborne, and orbital platforms; consequent improvements in the scope, resolution, and quality of models and modeling efforts; the initiation of major new climate sensors and observational efforts that are now beginning to yield results for integrated scientific study, and the potential loss or degradation of others; and the completion of a variety of important assessments, including the IPCC Fourth Assessment Report, assessments by WMO/UNEP, the Arctic Climate Impact Assessment, and CCSP's own synthesis and assessment products. As original questions are answered, new questions emerge, approaches evolve, and new users for CCSP's information step forward.

As a direct result of the past 4 years of program activity and progress, as well as recognition of the important changes to Earth systems noted above, there are significant new demands on CCSP. The most substantial of these is the need mentioned above for information at regional to local scales that are pertinent to direct land and resource management issues, in order to support decisionmaking. The development of robust partnerships (e.g., with state and local governments, academia, industry, public utilities, and nongovernmental organizations) will be an essential component of CCSP's response to these needs. These areas include not just climate change itself but improved understanding of associated issues of climate change impacts, adaptation, vulnerability, and sustainability, as well as the need for tools for the delivery of information for decision support in a manner that is both timely and useful, and at scales that are relevant to stakeholders' needs.

The previous sections of this report are intended to provide an overview of the structure and purpose of CCSP, its products, accomplishments, and challenges, and the progress that has led to the emergence of new priorities and changed emphases over the past 4 years. The remainder of this document is devoted to the articulation of plans for CCSP both programmatically and as related to CCSP's strategic goals for the period 2008 to 2010.

V Climate Change Science Program Cross-Cutting Elements

V. Climate Change Science Program Cross-Cutting Elements

In addition to its research plans aimed at achieving objectives associated directly with CCSP's strategic goals, CCSP has a series of cross-cutting elements that bridge research goals and either contribute to, or promulgate, its research accomplishments. CCSP will continue to explore ways in which to improve and extend progress toward achieving these cross-cutting programmatic elements. Issues related to the cross-cutting elements of observations systems and networks, modeling, decision support/stakeholder engagement, and communication of CCSP results to the public, nongovernmental organizations, the climate change technology community, and state and local officials and other decisionmakers are among the areas for needed growth that were identified by stakeholders and advisory committees, including the NRC in its recent report on CCSP progress (NRC, 2007a). Over the next 3 years, as a part of its long-range strategic planning, CCSP will actively consider approaches to determine and implement effective responses to these needs.

Observations and Data Management

Observation and monitoring are fundamental to understanding how the climate system works, how it is changing, what the uncertainties are, and the ways in which climate and other Earth systems are linked. As shown in the preceding discussion of research progress, observations and monitoring data from a variety of spaceborne, airborne, and *in situ* platforms provide critical data for understanding a myriad of climate parameters (e.g., external forcings like solar radiation, atmospheric processes such as the formation of ice in clouds, land surface processes like hydrology or the sensitivity of ecosystems to change, and ocean processes including the uptake of carbon). Without comprehensive data that provide measurements of essential climate variables through time and around the world, existing and future progress in understanding and forecasting the climate system and its interactions with and feedbacks to other Earth systems and human systems would not be possible.

CCSP provides active stewardship of observations that document the evolving state of the climate system, allow for improved understanding of its changes, and contribute to improved predictive capability for society. Some of these observations are not part of the CCSP budget (e.g., operational satellites) but are crucial to its success. A core CCSP activity is U.S. participation in the broad-based strategy of the international Global Climate Observing System (GCOS) in monitoring atmospheric, oceanic, and terrestrial domains with appropriate balance and integration of *in situ* and remotely sensed observations (for further information on GCOS activities and essential climate variables see <www.wmo.ch/pages/prog/gcos>). In recent years, observing activities by CCSP agencies have included the development and use of a number of platforms, experiments, and networks to measure and understand a wide variety of essential climate variables. These measurements range from satellite and airborne observations of atmospheric constituents, radiation and CO_2 fluxes, land- and sea-based carbon sources and sinks, rainfall, global land cover, ice, and snow, to ground-based networks and experiments measuring streamflow, water quality, soil moisture and composition, paleoclimate proxies, biomass and species distribution, and many others. All of these observations provide the essential underpinning for understanding Earth system processes, variability, and trends.

An area of recent concern for CCSP has been the ongoing declines in observational capabilities upon which the program relies. According to the NRC (2007c), "between 2006 and the end of the decade, the number of operating [remote-sensing] missions will decrease dramatically and the number of operating sensors and instruments on NASA spacecraft, most of which are well past their nominal lifetimes, will decrease by some 40 percent." These include a gap in continuous high-quality observations of land-cover change from Landsat and

similar sensors, and demanifestation of several climate sensors from the National Polar-orbiting Operational Environmental Satellite System (NPOESS) (some of which have now been restored, including high vertical profile ozone measurements and Earth radiation budget measurements on the NPOESS preparatory project mission).

Landsat-like sensors are required to systematically measure trends in global land cover to better understand land-climate interactions and to maintain ecosystem goods and services for society, and an interim solution is needed to provide this type of data as aging Landsat platforms fail and before planned follow-on capabilities are launched. Endeavors like the Mid-Decadal Survey – the goal of which is to develop a global, orthorectified data set from Landsat or Landsat-like observations based on measurements during 2004 to 2006 – will provide some continuity of these observations for research topics like global land cover, but other global change research that needs repeated data over short time spans (e.g., regional studies of phenological changes) will be affected. Similarly, past declines in ground-based monitoring networks like stream gages have hampered the Nation's ability to assess trends in important global change variables. Another area of CCSP focus is on upgrading measurement systems needed to support research that investigates linkages of clouds and atmospheric aerosols to climate processes, two high-priority science questions. Ensuring the long-term integrity and understandability of data products provided by important remote-sensing and *in situ* observing and monitoring systems will be a continuing challenge for CCSP agencies.

Over the next 3 years, observing activities by CCSP agencies will include foci on observing the polar climate as part of the International Polar Year (IPY) and developing the records needed for the Mid-Decadal Survey. The IPY plans to advance polar observations by establishing a variety of new multidisciplinary observatories using sensor web technologies (networks of spatially distributed sensor platforms that wirelessly communicate with each other) and power-efficient designs. Data from these, as well as more traditional surface- and space-based observatories, will provide the initial long-term, high-quality sustained measurements needed to detect future climate change. For the IPY and beyond, the United States plans to increase its efforts on observations of the polar atmosphere, ice, and ocean, as well as leverage its investments in polar research with international partners.

Data management and distribution activities, including those mentioned in Appendix 2 as examples, play a key role in making accessible the information necessary to fulfill CCSP's mission to provide the "Nation and the global community with the science-based knowledge to manage the risks and opportunities of change in the climate and related environmental systems." Many measurement and monitoring technologies and derived data systems benefit from the ongoing research and development under the aegis of CCSP, and from other Earth observation activities that are currently underway and in which CCSP participates. All such measurement and monitoring systems constitute an important component of and complement to the measurement and monitoring research and development portfolio of CCSP's sister program, the Climate Change Technology Program (CCTP). Complementary to CCSP's focus on providing a framework for Federal agencies to cooperate and coordinate climate change science, CCTP provides agencies with a coordinating framework for the development of climate change technology. For additional information on CCTP's mission and technology areas, including energy supply, end use, and infrastructure; CO_2 capture and sequestration; non-CO_2 greenhouse gases; and its measurement and monitoring research and development activities, see <www.climatetechnology.gov>. Since aspects of these areas are of importance to climate science as well, it is anticipated that over the next several years, the two programs (CCSP and CCTP) will take advantage of opportunities to develop closer collaborations on research areas of mutual interest.

Cooperative efforts by CCSP agencies are moving toward providing an integrated and more easily accessed Earth information system that will effectively preserve, extend, and distribute information about the evolving state of the Earth. A few representative examples of specific agency efforts in data management are described in Appendix 2; these include the Earth Science Research, Education, and Applications Solutions Network; the Global Change Master Directory; the Earth Observing System Data and Information System; the Distributed Active Archive Centers; and the Carbon Dioxide Information Analysis Center. Although each activity has a single lead agency, participation involves many CCSP agencies, as well as state, local, and nongovernmental partners. See also CCSP's annual report to Congress (CCSP, 2007b, 2008) for further information on these and other observation missions, networks, and data management activities.

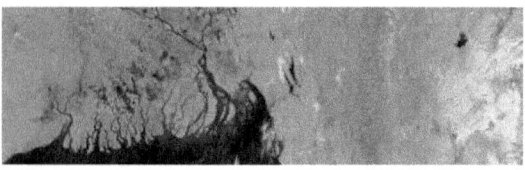

Modeling

Modeling is an essential tool for combining observations, theory, and experimental results, and using them to investigate and understand how climate and other components of the Earth system work, how they are affected by human activities, and how they will respond to potential future changes. The accuracy of models can be tested by comparing historic observational data and paleoclimate reconstructions with model results, providing that prior forcings are understood or can be estimated through proxy data. Models are vital tools for quantitatively integrating scientific understanding of the many components of the climate system. They are the only tools available for forecasting potential future changes. Comprehensive Earth system models are the primary tools for assessing the impacts of different future climate scenarios; they aim to represent the major components of the climate system (atmosphere, oceans, land surface, cryosphere, and biosphere) and to include the transfer of water, energy, chemicals, and mass among them, in order to understand how these factors and components interact. While much has been accomplished in the development of comprehensive climate models, much remains to be done. One factor limiting progress has been the paucity of data needed as inputs to the models, an area that CCSP has been addressing and continues to address in research under Goals 1 through 3 as discussed above. Another is that comprehensive climate models are complex and computationally intensive: they need an infrastructure designed to support their development and use, they require skilled personnel to design, implement, and improve them, and they require exceptionally large amounts of computer resources to run.

It is the need for modeling Earth systems of increasing complexity – incorporating the major relevant processes with the spatial detail and reliability sufficient to support decisionmaking – that drives the requirement for computational capabilities. Early climate models concentrated on simulation and prediction of the physical climate system. Over the past few years, CCSP modeling has begun to expand beyond the physical climate system to include complex models and modeling ensembles to elucidate the complex interrelationships of a fuller array of processes that make up the Earth system, including dynamics of ecosystems, biogeochemistry, and human influences on climate and responses to climate variability and change.

Beyond the increasing inclusion of multiple system elements, the most urgent need that has emerged in the modeling field is the need to move from models that produce reliable results only at global and continental scales to models that adequately address the regional and even local scales at which most societal, environmental, and land and resource management decisions are made. Incorporating new processes (e.g., carbon-climate interactions, aerosol-climate interactions), improving model physics to incorporate more complex system components and new parameterizations, and achieving greater fidelity of model simulations at a regional scale are CCSP's primary foci in modeling activities over the next 3 years. Modeling efforts to address specific aspects of modeling science and applications are underway in many CCSP member agencies, consistent with agency and CCSP missions. These efforts cover a wide range of modeling types and needs, from improvements in and refinements of global climate models to the development of regionally resolved decision-support models for land and resource management.

International Research and Cooperation

Since climate varies over a wide range of geographic scales that transcend national boundaries, climate change and its impacts are inherently international in scope. The study of climate change and variability on appropriate scales thus requires international cooperation among scientists, research institutions, governments, and intergovernmental agencies. CCSP supports and encourages international cooperation to increase leverage of resources, to enhance global observational capabilities and networks, to ensure data quality and exchange, and to foster the development of new scientific capabilities and applications in both the developed and developing countries of the world.

The United States – through CCSP, individual agency, and multi-agency efforts – participates in and supports a wide range of international cooperative activities related to global change and climate change research. These activities include support of key international climate change science research programs, especially those under the aegis of ICSU, and their regular review; support of ongoing international assessments; support of regional global change research networks; participation in informal organizations that foster global change research; and support of international efforts aimed at improving and coordinating observations of Earth. CCSP gives high priority to participation in international partnerships including IGBP, the International Human Dimensions Programme (IHDP), WCRP, DIVERSITAS, and the Earth System Science Partnership (ESSP).

Individually, CCSP-participating agencies support international activities that correspond with specific agency goals or missions, and/or for which they have been given the lead for the Federal government. In the latter case, and where appropriate, CCSP also provides a centralized structure for soliciting, communicating, coordinating, compiling, and transmitting U.S. input to a variety of international organizations addressing climate and global change research issues of importance to the United States. This includes support to the Department of State regarding programs of the United Nations specialized agencies involved in climate and global change research, IPCC, UNFCCC, and bilateral agreements for cooperation in climate change science and technology.

CCSP provides appropriate U.S. shares of multilateral funding for centralized coordination of international research programs that are important to international cooperation. This centralized coordination allows the international programs to collaborate with national research networks on disciplinary and interdisciplinary scientific endeavors and allows for the coordination of important synthesis reports. CCSP provides financial support to activities of the IPCC Working Group I Technical Support Unit and the partner programs of ESSP including the SysTem for Analysis, Research and Training (START). Long-term and active participation in and contributions to regional research networks including the Inter-American Institute for Global Change Research (IAI), the Asia Pacific Network for Global Change Research (APN), and burgeoning African cooperation fosters global change research in developing countries, develops research-driven capacity in those countries, and fosters research partnerships that ultimately support global goals in

research into and observations of the Earth system. The United States, through CCSP, also participates in a variety of programs that promote cooperation with other countries, directly enhance research capabilities in developing countries, and enhance climate forecasting and thereby adaptive capacity to respond to climate change in developing countries. More detail on the international partnerships and programs that CCSP and participating agencies support can be found each year in CCSP's annual report to Congress.

Decision Support

CCSP sponsors and conducts research that is ultimately related to policy and adaptive management decisionmaking. CCSP's decision-support approach is guided by several general principles, including the early and continuing involvement of stakeholders, the explicit treatment of uncertainties, the transparent public review of analysis questions, methods, and draft results, and the evaluation of lessons learned from ongoing and prior decision-support and assessment activities. CCSP utilizes three approaches to decision support: 1) assessments, including CCSP's synthesis and assessment products; 2) adaptive management; and 3) policy-relevant analyses. These approaches inform the integrated decision-support activities of the program and are also utilized as appropriate by participating agencies in their decision-support activities.

A key set of current CCSP decision-support results in preparation at this time are its 21 synthesis and assessment products that integrate research results focused on key issues and related questions frequently raised by decisionmakers (see Appendix 1). Up-to-date evaluations of current science can be used for informing public debate, policy development, and adaptive management decisions and for defining and setting the future direction and priorities of the program. The synthesis and assessment products constitute an important new form of topic-driven integration of U.S. global change assessment efforts. These CCSP products are U.S. Government reports, subject to the provisions of the Information Quality Act[5]

---

[5] Section 515 of the Treasury and General Government Appropriations Act of 2001.

and the Federal Advisory Committee Act Amendments of 1997.[6] The synthesis and assessment products are generated by researchers in a process that involves review by experts, public comment from stakeholders and the general public, and final approval by the departments/agencies that participate in CCSP. A detailed description of the process involved in the preparation of these reports can be found in *Our Changing Planet: The U.S. Climate Change Science Program for Fiscal Year 2008*.[7] Up-to-date information on the preparation and publication status of all SAPs is available from <www.climatescience.gov/Library/sap>.

To build on the experiences of earlier assessment activities, CCSP requested that the NRC carry out an analysis of global change assessments that have addressed topics broadly similar to those encompassed by CCSP. The study, which was released in early 2007, included a comparative analysis of past assessments that address issues directly related to the science and technical issues of CCSP (NRC, 2007b). The committee concluded that global change assessments are extremely important for informing decisionmakers. In identifying essential properties of a successful assessment, it stressed that future assessment processes must communicate relevant information to the user, address the technical quality of the information, and demonstrate fairness and impartiality in the assessment process. The report identifies a number of essential elements that increase the probability that an assessment will effectively inform decisionmakers and other target audiences. CCSP will build the findings of the NRC into its future assessment activities.

Education

Many of the CCSP agencies have a substantial investment and interest in climate education. These activities include the development of educational frameworks, classroom materials, and fact sheets for students and the public; provision of up-to-date information on climate science via teacher workshops, web sites, podcasts, RSS feeds, lectures, and many other venues; and funding of informal education activities (e.g., museums). CCSP is exploring ways to

enhance and better coordinate and integrate these activities. CCSP will also explore opportunities for partnerships with educational organizations that will help it capitalize on its investments and aid in the development of increased climate literacy for the Nation.

Stakeholder Engagement, Communication, and Dissemination of Results

CCSP agencies will continue to take a leadership role in the dissemination of results and products that come from the program's research, observations, and decision-support activities. The premier dissemination method for the near term is the publication of CCSP's synthesis and assessment products aimed at a range of audiences including scientists, policymakers, and land and resource managers. The program will ensure that the conclusions from its assessment products and activities are widely communicated. In addition, the program will coordinate the development of interagency climate-related communications with those of the member agencies to help ensure that the accomplishments of the overall national investment in climate-related science are understood and are widely available to potential users.

Individual scientists and science teams continue to publish research in the open, peer-reviewed scientific literature, and this is the most prolific source of climate change science information produced under the aegis of the program, with hundreds of scientific papers written, reviewed, and published each year. CCSP will continue to develop and hold stakeholder workshops for mutual education and information exchange.

A number of public comments received during the preparation of this Revised Research Plan highlighted the need for research on how to effectively disseminate climate change prevention and adaptation information to key professional and public audiences, and made the suggestion that CCSP should also develop social marketing and behavior change research that examines how to move members of key audiences beyond awareness to taking effective action, akin to successful efforts undertaken in the public health sector. This area of research will be given attention as part of CCSP's strategic planning process.

---

[6] P.L. 105-153, Sec. 2(A), (B), Dec. 17, 1997, 111 Stat. 2689.
[7] See <www.usgcrp.gov/usgcrp/Library/ocp2008>.

VI

Climate Change Science Program Research Plans: 2008 to 2010

VI. Climate Change Science Program Research Plans: 2008 to 2010

The scope of CCSP scientific research is far-reaching. CCSP strategic goals encompass everything from basic scientific research on Earth's past and present climate and climate variability, the forces that result in changes in Earth's climate and related systems, reducing uncertainties in projecting future change and its consequences, and the sensitivity/adaptability of both ecosystems and human systems, all the way to the application of the knowledge gained to the decisionmaking process for the management of risks and development of strategies for adaptation to climate change.

Over the past several years, some scientific questions have been answered, some remain or have evolved, and new questions have emerged as a consequence of the progress made. For example, 4 years ago a key question identified in CCSP's Strategic Plan was whether there is a cause and effect relationship between human activities and a warming climate. Today, as reflected in the IPCC Fourth Assessment Report and largely due to the valuable research fostered by CCSP and other programs worldwide, that question has been answered in the affirmative. Similarly, at the time of the IPCC Third Assessment Report, there was great uncertainty regarding whether melting of the Greenland and Antarctic ice sheets could contribute to sea-level rise; in the Fourth Assessment Report, the data show a high likelihood that there was a contribution over the period 1993 to 2003. In both these cases, the question is no longer whether such changes are occurring, but rather how much, how fast, and where the consequences of these changes will be felt.

A third and slightly different example is in the field of climate modeling. Over the past several years, advances in the state of knowledge of model physics together with the rapidly increasing capacity of computational resources mean that it is now possible to begin to develop scenarios showing potential regional, rather than continental or global, changes and impacts. Examples of modeling efforts that are providing value at regional to local scales are addressed in the research plan discussions that follow.

Overall, increasing lengths and quality of long-term monitoring records, and their increased availability to researchers via electronic means, provide greater resolution and reduced uncertainty in understanding trends across a wide variety of scientific endeavors. These developments have led overall to an increased demand for sound science to support decisions, and an emphasis on the relevance of science to questions of adaptation to and mitigation of climate change. Thus, the Nation's ongoing investments and advances in basic scientific monitoring, observations, process research, and modeling remain vital to global change research.

CCSP will utilize the Nation's investments and advances in monitoring and measurement capabilities to continue to gather and analyze information via experimental manipulation, measurement, modeling, and assessment studies to enhance understanding of Earth systems and the processes affecting their changes (Goals 1 through 3), while moving into increased activity in understanding societal and ecosystem sensitivities, predicting environmental responses, and supporting decisionmaking (Goals 4 and 5). Research is needed to more fully understand the implications of climate change for both natural and managed ecosystems and to strengthen the delivery of that information to land and resource managers and other stakeholders to support adaptive management and mitigation efforts. Also needed are improvements in the reliability of ecological forecasting, and the development of better capabilities and an early warning system for anticipating potential abrupt climate changes and climate change impacts. Research is needed to understand the human health implications of global change and the potential impacts of global change on infrastructure, economic, and other human systems so that risks can be managed and vulnerability reduced.

Comments received during the public comment period and expert review of this Revised Research Plan highlighted the need for improved knowledge through research on social, economic, behavioral, engineering, and multidisciplinary research across the natural/physical science and socioeconomic science divide, as well as research on dissemination and information transfer. The comments suggested that perhaps one of the near-term goals for the CCSP should be to develop a more extensive and inclusive socioeconomic research agenda to meet its goals. These areas of research will be given attention as part of CCSP's strategic planning process. Efforts are underway to begin to strengthen CCSP's activities in these areas, through the active participation of agencies with missions in these areas, through direction to the IWGs to address and include Goals 4 and 5 in their collaborative planning, and through planning for both near- and long-term engagement with and inclusion of a wide range of stakeholders through such venues as workshops, listening sessions, and participation in mutually relevant professional meetings.

As stated in Section I, CCSP's strategic goals provide the focus and direction for the program, to ensure that knowledge developed by the participating agencies and research elements can be integrated and synthesized, and this remains the overarching strategy for the program. The strategic goals are long-term goals that speak to major advancements in our understanding of Earth's climate, its variability, the factors that control and change it, and the consequences and impacts of those changes on Earth's systems and human communities. While the scientific enterprise must be able to respond quickly to make use of and understand unexpected events and results (e.g., recent observations of rapid change at high latitudes), the accomplishment of such crucial yet broad-ranging goals is often done incrementally, through multiple projects aimed at understanding specific questions, the results of which are combined to provide progress towards the overarching goal.

The following goal-by-goal descriptions provide a sense of the strategic purpose and scope encompassed by these goals, and the way in which the goals inform monitoring and research, decision support, and communications throughout the program. Included with each goal description is a series of illustrative examples of specific activities planned for the period 2008 to 2010. In some cases, these activities have longer range components that will continue after the 2008 to 2010 time frame. This goal-by-goal discussion is followed by a series of interagency implementation plans that, through their multidisciplinary nature, inherently address multiple goals and require the combined efforts of multiple agencies. In addition to the examples specified here, other urgent research topics, yet to be determined, will undoubtedly emerge from ongoing progress, events, and societal needs.

Research Plans for Goal 1: 2008 to 2010

CCSP Goal 1: Improve knowledge of the Earth's past and present climate and environment, including its natural variability, and improve understanding of the causes of observed variability and change.

OVERVIEW

Climate conditions change significantly over the span of years, decades, and even longer time scales. CCSP research will improve understanding of natural oscillations in climate on time scales from weeks to centuries, including improving forecasts of the El Niño-Southern Oscillation (ENSO), a large-scale climate oscillation with implications for resource and disaster management. Research will continue to sharpen qualitative and quantitative understanding of climate extremes, and to what degree any changes in frequency or intensity lie outside the range of natural variability, through improved observations, analysis, and modeling. The program also will continue to expand and refine observations, monitoring, and data/ information system capabilities and increase confidence in understanding of how and why climate is changing. Fostering improved interactions and connectivity between research and ongoing operational measurements and activities continues to be another important aspect of the program's work.

KEY RESEARCH TOPICS

The 2003 CCSP Strategic Plan identifies some key research topics that must be addressed in order to answer the broad questions of the range of Earth's natural climate variability through time, and to improve understanding of the causes and consequences of observed change. Understanding couplings between the oceans, atmosphere, and land are central to this goal. Research outcomes identified in the Strategic Plan include:

- Better understand natural long-term cycles in climate (e.g., PAO and NAO)

- Improve and harness the capability to forecast El Niño/La Niña and other seasonal-to-interannual cycles of variability

- Sharpen understanding of climate extremes through improved observations, analysis, and modeling, and determine whether any changes in their frequency or intensity lie outside the range of natural variability

- Increase confidence in the understanding of how and why climate has changed

- Expand observations and data/information system capabilities.

These outcomes are important as a necessary step toward producing better forecasts of seasonal climate change and the regional structure of change associated with global warming, and uncertainty estimates for these forecasts. Understanding the complex connections between oceans, atmosphere, and land that feed into climate variability requires a diverse and extremely broad set of long-term research activities, including observation and monitoring efforts and modeling. These include the need for information on paleoclimate and ocean chemistry through time; basic information on sea surface temperature through time; measurements of ocean salinity and pH, carbon cycling, chemical properties and changes (including ocean acidification); physical properties and changes; biological abundance and variation in the world's oceans; basic atmospheric measurements worldwide; and the linkages of clouds and atmospheric aerosols to climate processes.

On the land surface, among the highest priorities are basic measurement and monitoring of rapidly changing high-latitude parameters and their connections to the climate system, including permafrost temperatures and carbon cycling, rates and distribution of melting of both northern and southern polar ice, the effects of Arctic aerosols on clouds, radiation budget, and ice melting, and a better understanding of land use effects on ecosystem processes and coupled land-climate interactions. Also key are the need for better understanding of ecological and biogeochemical processes and feedbacks and their relationships to climate, atmosphere, and hydrology, and the role of land use in these processes. These topics are among those addressed by CCSP's 2008 to 2010 research

plans as illustrated in the examples below. These are illustrative examples intended to provide some specificity regarding the program's breadth in each area, but are not intended to be exhaustive lists of plans.

ILLUSTRATIVE GOAL 1 PLANS

Oceans

Global Climate and Ocean Observations. **Priorities for** advancement of atmospheric and ocean observing components are carried out in significant part under the GCOS framework. These priorities include (1) reducing the uncertainty in the carbon inventory of the global ocean, sea-level change, and sea surface temperature; (2) continuing support for existing *in situ* atmospheric networks in developing nations; and (3) planning for surface and upper air GCOS reference observations consistent with CCSP Synthesis and Assessment Report 1.1. The Argo ocean observation array, which is making vertical measurements of ocean conditions, is reaching global coverage. The Tropical Atmosphere Ocean (TAO) array in the Pacific will begin to be refreshed with redesigned mooring technologies, and the TAO tropical system will be expanded further in the Indian Ocean. Three new ocean reference stations will be added to the system, for improved forecasts and modeling validation, assessments of climate impacts on ecosystems, and monitoring for possible rapid climate change. The tide gauge network will continue to be upgraded for real-time reporting, also contributing to the international tsunami warning system. Continued support will be given to the activities, database development, and data delivery systems of the international Global Sea Level Observing System. The Global Ocean Observing System will make incremental advances, building up to 62% completion: 50 surface drifters will be equipped with salinity sensors for satellite validation and salinity budget calculations, particularly in the polar regions; a new reference array will be added across the Atlantic basin to measure changes in the ocean's overturning circulation, an indicator of possible abrupt climate change; and a pilot U.S. coastal carbon observing network will enter sustained service, to help quantify North American carbon sources and sinks

and to measure ocean acidification caused by CO_2 sequestration in the ocean. Work is underway on developing biological sensors as part of the Global Ocean Observing System. Finally, planning activities will continue on development of a GCOS Reference Upper Air Network to aid in enhancing the quality of upper tropospheric and lower stratospheric water vapor measurements.

Field Experiment to Improve Understanding of Southeast Pacific Climate Processes. The Variability of the American Monsoon System (VAMOS) Ocean-Cloud-Atmosphere-Land Study – Regional Experiment (VOCALS-REx) is planned for October and November 2008, with data analysis and modeling to follow in FY 2009, FY 2010, and beyond. This international field experiment is designed to better understand physical and chemical processes central to the climate system of the Southeast Pacific (SEP) region. The climate system of the SEP involves tightly coupled, but poorly understood, interactions among the ocean, atmosphere, and land. VOCALS-REx will focus on interactions among clouds, aerosols, marine boundary layer processes, upper ocean dynamics and thermodynamics, coastal currents and upwelling, large-scale subsidence, and regional diurnal circulations to the west of the Andes mountain range. Multidisciplinary intensive observational data sets will be obtained during VOCALS-REx from several platforms including aircraft, research vessels, and a surface land site. An intensive observational period will take place during October and November 2008, when the extent of stratocumulus over the SEP is at its greatest, the southeast trade winds are at their strongest, and the coupling between the upper ocean and the lower atmosphere is at its tightest. This project is one of several that will contribute to the activities of the U.S. Climate Variability and Predictability (CLIVAR) Program, which is responsible for coordinating many of CCSP's activities pursuant to its physical science objectives under Goal 1.

Atmosphere

Tropical Composition, Clouds, and Climate Coupling. CCSP researchers will begin their analysis of data from a FY 2007 field mission in Costa Rica to study how climate is linked to atmospheric composition and clouds in the tropical summer convective wet season. This successful Tropical Composition Cloud and Climate Coupling field mission involved three major aircraft (DC-8, ER-2, WB-57F) making over 20 research flights using some 60 instruments, including balloon launches from three locations, and involving over 250 participants. Analyses will incorporate data from aircraft flights and ground

measurements, as well as Aura satellite observations, to address scientific questions related to how clouds, aerosols, and trace gases influence radiative heating in the very active tropical atmosphere.

International Polar Year Research on Arctic Aerosols and their Connections to Clouds, Radiation, and Ice Melting. The long-range transport of anthropogenic pollution from North America, Europe, and western Asia creates the aerosols associated with the so-called Arctic haze, a phenomenon that recurs every winter and spring. The direct and indirect climate impacts of the aerosols can be quite different in the Arctic compared to elsewhere, because high surface reflections from snow and ice mean that even weakly absorbing aerosol layers can heat the Earth/atmosphere system in the Arctic. The analysis of 8 years of Atmospheric Radiation Measurement (ARM) data suggests that enhanced aerosol concentrations may increase the amount of thermal energy emitted by many Arctic clouds to the surface, causing increased Arctic warming in addition to the greenhouse gas warming (Lubin and Vogelmann, 2006).

As part of IPY research, CCSP scientists will conduct field missions to investigate Arctic aerosol/climate connections in this unique environment. Spring and summer measurements from satellites, aircraft, and the surface will be made in collaboration with the larger IPY study POLARCAT (Polar Study using Aircraft, Remote Sensing, Surface Measurements, and Models of Climate, Chemistry, Aerosols, and Transport) which is a core IPY activity focusing on the transport of aerosols and pollution from mid-latitude anthropogenic sources and boreal fires to the Arctic and its climate impact (<www.polarcat.no/>). The Arctic Research of the Composition of Troposphere from Aircraft and Satellites (ARCTAS) experiment will also perform comprehensive measurements of trace gases, aerosols, and radiation from three aircraft during two seasons (see <www.espo.nasa.gov/arctas/>). Spring flights were focused on the climate impact of Arctic haze and anthropogenic pollution transported to the Arctic from mid-latitudes, and observations were made to assess the long-range transport of anthropogenic pollution to the Arctic and its contribution to Arctic haze, ozone chemistry, and the possible connections between Arctic aerosols and the melting of polar ice through the Aerosol, Radiation and Cloud Processes affecting Arctic Climate campaign and the Indirect and Semi-Direct Aerosol Campaign (see <www.esrl.noaa.gov/csd/arcpac/> and <acrf-campaign.arm.gov/isdac/>). Summertime observations will be made to assess fire emissions from the boreal forests. ARCTAS flights are being closely coordinated with satellite observations from instruments on the Terra and A-Train satellites with the goal of improving their long-term use in diagnosing changes in Arctic atmospheric conditions. Ongoing and future analyses of these measurements will ultimately improve the ability of current models to simulate the influence of anthropogenic pollution and boreal fires on the Arctic atmosphere and climate.

Land

Prototype Land-Cover Mapping Activities. The National Land Cover Database effort in Alaska, Hawaii, and Puerto Rico was finished in December 2007, marking completion of the first compilation of nationwide land cover ever produced at 30-m resolution. This will improve knowledge of Earth's present environment and its variability and form the baseline for quantifying change at high spatial resolution.

Development of a Permafrost Monitoring Network. Climate projections by coupled atmosphere-ocean general circulation models suggest significant environmental changes will occur in the Arctic during the next 80 years. Given the large potential impacts, and the significant uncertainty in the model projections, the U.S. Department of the Interior is developing a long-term permafrost monitoring network on Federal lands in northern Alaska; this network contributes to the Global Terrestrial Network for Permafrost and GCOS. Analysis of data acquired thus far by the monitoring network suggests that permafrost temperatures on the western half of the Arctic Coastal Plain in Alaska may have warmed several degrees Celsius between 1980 and 2005.

Land-Oceans-Atmosphere Integration (Climate System)

Polar Region Observations: International Polar Year. Integrated polar climate observations will continue to be a CCSP focus. As a part of IPY, CCSP research will investigate the possible connections between Arctic haze aerosols and the melting of polar ice in the region. The investigation will involve multiple agencies in cooperation with scientists and facilities from several other countries. In addition to a wide variety of surface measurements, *in situ* and remote-sensing measurements will be made from balloons and aircraft. Satellite observations will include Cloud-Aerosol Lidar and Infrared Pathfinder Satellite Observation (CALIPSO) and Cloudsat, using lidar and radar instruments to provide three-dimensional distributions of aerosols and layered clouds. These data – when combined with data from the A-train configuration of the Aqua, Aura, and

Parasol satellites orbiting in formation – will enable systematic observation of the key climate forcing of aerosol indirect effects, climate sensitivity of cloud feedbacks, and polar climate response of difficult-to-observe polar clouds. Surface field teams from many nations will provide important *in situ* measurements and analyses to complement the satellite observations contributed by multiple space agencies. A U.S. Climate Reference Network system will be deployed at the Russian Arctic site of Tiksi at latitude 71.5°N in order to provide long-term reference measurements of temperature, precipitation, wind, pressure, and surface radiation in support of IPY and beyond. These efforts are tied to those of the Arctic Observing Network (AON), which is envisioned as a system of atmospheric, land, and ocean-based environmental monitoring capabilities – from ocean buoys to satellites – that will significantly advance observations of Arctic environmental conditions. Developed largely as a research system, it is hoped that data from AON will eventually enable the interagency U.S. Government initiative – the Study of Environmental Arctic Change – to better understand the wide-ranging series of significant and rapid changes occurring in the Arctic.

Data Fusion. As the length of record in the database of global observations increases, effort will be placed on developing methodologies for assimilating Earth observations into global climate models, to produce an integrated view of the climate system and to better provide this view to users as part of decision support and resource management systems. The value of the data will benefit by increased "data fusion" in which, for example, MODIS observations will be joined with complementary capabilities of other Earth-observing instruments to provide much improved, more accurate and rigorous observations of key phenomena such as sea surface temperature, cloud characteristics, and land surface features. Data fusion efforts will include instruments on existing Earth Observing System (EOS) missions such as Terra, Aqua, ICESat, Aura, Landsat, and SORCE; those on recently launched missions such as CloudSat and CALIPSO; and, farther in the future, those on Earth System Science Pathfinder missions and the Global Precipitation Measurement Mission. The fusion of space-borne observations with *in situ* biological and physical observations, such as those gathered through the Ameriflux Network and the Ocean Observing Initiative, is crucial for gaining a better understanding of trends in and associated consequences of the variability in the atmosphere-land-ocean system. This activity is closely related to the CCSP

Climate Variability and Change research element's priority of improving Earth system analysis capabilities.

Continuity of Climate Measurements. As new satellite instruments bring new measurement capabilities, the challenge becomes establishing priorities for the right mix of existing observing capabilities and new capabilities to support the goals of CCSP. Continuity of measurement of several key climate variables is being carefully considered, including stratospheric ozone, radiative energy fluxes of the Sun and Earth, atmospheric CO_2 and methane concentrations, global surface temperature, and global land cover (e.g., as measured by Landsat).

The long-term record of global land cover was begun by Landsat 1 in 1972 and continues through the collection of data from Landsats 5 and 7. Launched in 1984, with a design life of 3 years, Landsat 5 continues to provide near-global coverage through a network of international ground station cooperators. Landsat 7 was launched in 1999, and continues to acquire global observations on a daily basis although in a degraded operating mode. The combined assets of Landsats 5 and 7 permit repeat coverage as frequently as every 8 days over ground-receiving station sites. Efforts to create a long-term record of global land cover, started by Landsat in the 1970s, are currently underway for the transition to a Landsat Data Continuity Mission (LDCM) being planned by NASA and the U.S. Geological Survey (USGS). LDCM is expected to have a 5-year mission life, with 10-year expendable provisions. The National Land Imaging Program Plan, which is currently proposed for FY 2009, will aid in long-term planning for a stable, operational, space-based land imaging capability.

Planning continues on deploying component sensors from NPOESS. Several of the climate-related instruments that had been demanifested from NPOESS have now been restored, including high vertical profile ozone measurements and Earth radiation budget measurements on the NPOESS preparatory project mission, development of Clouds and the Earth's Radiant Energy System (CERES) and the Total Solar Irradiance Sensor (TSIS) on other operational platforms, and planning for the 2013 launch of Jason-3. The NPOESS Preparatory Project is scheduled for 2010 as a bridge mission between NASA's EOS program and NPOESS, now scheduled for its first launch in 2013.

The record of precipitation that has been extended in recent years to include oceanic as well as land areas using measurements from the Tropical Rainfall Measuring

Mission (TRMM) is another example of a key climate data set that needs to be considered as priorities are set for the future. These examples of key climate variables are elements of the comprehensive observing system to monitor changes in the cycles of carbon, energy, water, and related biogeochemical processes that drive Earth's climate.

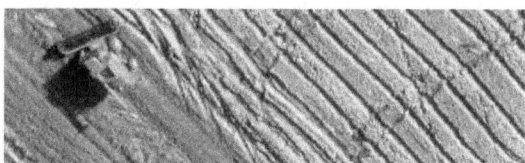

Research Plans for Goal 2: 2008 to 2010

CCSP Goal 2: Improve quantification of the forces bringing about changes in the Earth's climate and related systems.

OVERVIEW

Forcings are factors that directly change the average energy balance of the Earth-atmosphere system by affecting the balance between incoming solar radiation and outgoing or "back" radiation. They include external factors like variations in solar radiation and volcanic emissions, and human-induced factors such as combustion of fossil fuels, changes in land cover and land use, and industrial activities. Variations in solar radiation directly affect the amount of energy received by Earth, and volcanic emissions and human-induced forcings produce greenhouse gases (GHGs) and aerosols that alter the composition of the atmosphere and affect the physical and biological properties of the Earth's surface. These changes have several important climatic effects, the quantification of which has improved dramatically in recent years but upon which a substantial amount of work remains to be done. Research conducted through CCSP will continue to address the reduction of uncertainty in the sources and sinks of GHGs and aerosols and their precursors; the long-range atmospheric transport of GHGs and aerosols and their precursors; and the interactions of GHGs and aerosols with global climate, ozone in the upper and lower layers of the atmosphere, and regional scale air quality. This research will continue to improve quantification of the interactions and feedbacks among the carbon cycle, other biological, biogeochemical, and ecological processes in terrestrial, freshwater, and marine systems, and land cover and land use to better project atmospheric concentrations of key GHGs and to support improved decisionmaking. The program will also continue to work toward improved capabilities for developing and applying emissions scenarios in research and analysis, in cooperation with its sister program CCTP.

KEY RESEARCH TOPICS

The 2003 CCSP Strategic Plan articulates the research questions that must be addressed in order to improve understanding of the forces that are responsible for changes in the Earth's climate and related systems. Understanding solar radiation, the physics of the atmosphere, the effects of atmospheric trace constituents and aerosols that change the properties of Earth's atmosphere, and the movement of carbon through ocean, atmosphere, and land systems is crucial to this endeavor. For the land surface, a key uncertainty is the potential effects and feedbacks of ecological change (e.g., changes to land cover that in turn cause changes in the reflectivity of the land surface) on the climate system. Research outcomes identified in the CCSP Strategic Plan include:

- Reduce uncertainties about the sources and sinks of GHGs, emissions of aerosols and their precursors, and their climate effects

- Monitor the recovery of the ozone layer and improve the understanding of the interactions of climate change, ozone depletion, tropospheric pollution, and other atmospheric issues

- Increase knowledge of the interactions among emissions, long-range atmospheric transport, and transformations of atmospheric pollutants, and their response to air quality management strategies

- Develop information on the carbon cycle, land cover and use, and biological/ecological processes by helping to quantify net emissions of CO_2, methane, and other GHGs, thereby improving the evaluation of carbon sequestration strategies and alternative response options

- Improve capabilities to develop and apply emissions and related scenarios for conducting "If…, then…" analyses in cooperation with CCTP.

Understanding the forces that act upon climate and related Earth systems requires an integrated and broad-based approach, to understand how diverse and various changes in the land, oceans, and atmosphere will affect and potentially feed back into the climate system. This calls for a detailed knowledge of the physics of atmospheric

processes and constituents including the abundance and variations in trace species (including non-CO_2 carbon-bearing constituents like methane and pollutants like hydrofluorocarbons) and aerosols (including dust particles and particulate emissions), and how these processes and constituents vary through time and affect cloud formation, radiation balance, and feedbacks among the land, oceans, and atmosphere. In order to understand climate forcings, it is also necessary to understand the land surface changes that affect the transfer of radiation from atmosphere to land and back, and the processes that control the cycling of carbon on land, in the world's oceans, and from land surfaces to coastal waters and beyond. A particular area of urgent need is for this information at Earth's high latitudes, where changes are occurring rapidly. Understanding these forcings and feedbacks is integral to building scenarios for understanding the role of emissions in climate. The 2008 to 2010 research topics outlined below are among those that address these priorities. These examples are intended to be illustrative of the breadth of the program's activities under Goal 2, but are not intended to be exhaustive lists of the program's implementation of its strategic directions described above.

ILLUSTRATIVE GOAL 2 PLANS

Oceans

Global Ocean Carbon. Ongoing and new studies will continue to improve understanding of the ocean carbon cycle and its effects on ocean carbon dynamics. Of particular interest are the feedbacks and drivers of ocean chemistry and biology that affect carbon uptake and sequestration by the ocean, including the biotic and abiotic partitioning of carbon, and constraints on ocean carbon sequestration. An interagency field study of air-sea CO_2 flux and remotely sensed data in the high-latitude Southern Ocean during the 2007-2008 austral summer focused on understanding both (a) the kinetics of gas exchange and the factors controlling it, and (b) the physical and biogeochemical factors controlling the exchange of CO_2 across the air-sea interface, in the context of developing parameterizations for those factors that can ultimately be remotely sensed to determine regional and global air-sea CO_2 fluxes. The Southern Ocean Gas Exchange Experiment (SO GasEx) is being conducted in 2008 in the Atlantic sector of the Southern Ocean. Shipboard studies include physical, chemical, biological, and meteorological measurements. Analyses of the data collected during the field campaigns will continue through 2008 and 2009.

Atmosphere

Atmospheric Monitoring. Measurements of radiative trace species (including CO_2 and methane), started 6 years ago at Summit, Greenland, will continue, and measurements will be expanded to include tracers of carbon sources (e.g., hydrofluorocarbons) with new instrumentation. Weekly carbon cycle flask measurements will continue across Arctic areas in Canada, Norway, Iceland, Finland, the North Atlantic, and Alaska. A collaborative effort of U.S. agencies in the Yukon will install instrumentation for continuous vertical sampling within a 300-m boundary layer. Aircraft sampling for carbon cycle gases will be expanded to sites in Saskatchewan, Canada, and Poker Flats, Alaska. Where possible, aircraft sampling will be added over Churchill on Hudson Bay, Manitoba, and several other sites. International cooperation continues with Russia at the Baseline Observatory on the central Siberian Arctic Ocean coast where carbon gas measurements will be conducted, particularly the measurements of methane that could be released from wetlands as the northern high latitudes continue to warm. In the Antarctic, flask gas sampling will continue at the South Pole Station, and at Halley, Syowa, and Palmer on the Antarctic coast. Carbon flask sampling will continue across the Drake Passage and around Antarctica on the support ship from Ushuaia to Palmer and the annual Chinese cruise, respectively.

The Ice in Clouds Experiment. The Ice in Clouds Experiment (ICE) took place in November 2007, and data analysis and interpretation are ongoing. The goal of this study is to improve understanding of ice nucleation in the atmosphere. This knowledge will improve the modeling of ice cloud formation, precipitation, and climate effects. The specific objective of ICE is to show that under given conditions, direct measurements of the thermodynamic and kinetic environments of clouds (temperature, humidity, wind) and specific measurable characteristics of the aerosol, including chemical composition, can be used to predict the number of tiny ice particles that are initially seeded by existing atmospheric particles. ICE used airborne measurements of clouds along with coordinating ground measurements in mountainous locations such as the Front Range of Colorado and Wyoming. Close collaboration between theory, field, lab, and modeling studies are emphasized and ongoing.

Application of the ARM Mobile Facility to Study the Aerosol Indirect Effects in China. China has exceptionally high aerosol loading with diverse properties whose influence has been detected across the Pacific Rim. Because of the rapid pace of changes in the atmospheric environment over China, it provides a natural test bed for identifying and quantifying the climatic effects of aerosols. Preliminary analyses of multiple satellite data sets (MODIS and the TRMM Tropical Microwave Imager) indicate more complex and unique aerosol indirect effects than are found in relatively cleaner environments. Unfortunately, China is one of the least observed regions, especially in terms of aerosol and cloud properties. To this end, the U.S. Department of Energy (DOE) ARM Mobile Facility (AMF) is being deployed from 1 January to 31 December 2008 to investigate (1) the mechanisms of the aerosol indirect effects in the region and the roles of aerosols in affecting regional climate and atmospheric circulation with a special focus on the impact of the East Asian monsoon system, and (2) effects of long-range transport of aerosols to the Pacific Rim and the western United States. Analyses of data collected will follow in 2009 to 2010 and beyond.

Land

Yukon River Basin: An Arctic Benchmark. A developing consortium of U.S. and Canadian Federal, state, and provincial agencies, university scientists, and tribal organizations is initiating a major campaign to understand and predict climate-induced changes in the air, water, land, and biota within the Yukon River Basin (YRB). The consortium will implement a prototype environmental monitoring and research strategy that links air, water, soil, and forest information to understand sources, sinks, and uncertainties, and to characterize changes in carbon and energy budgets across the Arctic, boreal, and Arctic Ocean systems. This collaborative scientific campaign, using the YRB and adjacent coastal ocean as a representative landscape unit, will provide a benchmark for tracking and understanding changes occurring throughout the Arctic and subarctic region.

Networks of Observations. Systematic observations provide the basis for determining if there are changes in ecosystem processes and properties, and if there are indications that such changes can be attributed to climate, CO_2, or other factors. The AmeriFlux Network will continue to sustain a broad suite of observations at close to 100 U.S. sites that provide an important data record of CO_2 flux measurement, biotic responses to climate variables, meteorological conditions, and phenological trends that are believed to be associated with climate forcings. This unique and systematic set of observations is used increasingly for validating a wide range of ecosystem, landscape, and inverse model calculations. The network of observations also provides information for calibrating remotely sensed biotic and landscape observations. There also is a high degree of coordination among other observing networks from North America, Europe, and Asia, which collectively are building a powerful capability for understanding the fundamental biogeochemical and biophysical processes needed for more comprehensive Earth system models.

Land-Oceans-Atmosphere Integration (Climate System)

The North American Regional Climate Change Assessment Program. The North American Regional Climate Change Assessment Program (NARCCAP) is an effort funded by multiple agencies (NSF, DOE, and National Oceanic and Atmospheric Administration (NOAA)) to meet the climate scenario needs of the United States and Canada. NARCCAP aims to develop an ensemble of regional climate change scenarios for North America, and to develop and apply statistical methods to systematically investigate the uncertainties in future climate projections at the regional level. NARCCAP uses the multi-model approach to achieve its goal. This includes the use of four global atmosphere-ocean coupled models to provide global climate change simulations, the use of six regional climate models to downscale the global simulations, and the use of two global atmosphere models in time-slice experiments to provide climate change scenarios at the same horizontal resolution as the regional climate models. NARCCAP represents the first large-scale coordinated effort that includes international participation to develop and assess regional climate change scenarios for North America. Users of NARCCAP model outputs include university and Federal scientists and regional stakeholders. NARCCAP model archives are accessible from the Earth System Grid for analysis of regional phenomena, further downscaling to higher spatial resolution, and assessment of climate change impacts and uncertainties. A set of new high-resolution climate model simulations has been completed for North America that provides information at a scale finer than 100 km x 100 km (Han and Roads, 2004; Leung et al., 2004; Mason, 2004; Wood et al., 2004). The ability of these regional scale models to represent climate processes is being assessed (see <www.narccap.ucar.edu>). These regional and global simulations, based on models developed at U.S. institutions, contributed to the IPCC Fourth Assessment Report.

North American Carbon Program and Ocean Carbon and Climate Change Science and Integration. The work of the North American Carbon Program (NACP) is aimed at quantifying the magnitudes and distributions of terrestrial, freshwater, oceanic, and atmospheric carbon sources and sinks for North America and adjacent oceans; enhancing understanding of the processes controlling source and sink dynamics; and producing consistent analyses of North America's carbon budget that explain regional and continental contributions and year-to-year variability. This program is committed to reducing uncertainties related to the increase of CO_2 and methane in the atmosphere and the amount of carbon – including the fraction of fossil fuel carbon – being taken up by North America's ecosystems and adjacent oceans. The Ocean Carbon and Climate Change (OCCC) program is focused on oceanic monitoring and research aimed at determining how much atmospheric CO_2 is being taken up by the ocean at the present time and how climate change will affect the future behavior of the oceanic carbon sink. The terrestrial and ocean carbon programs are synergistic, integrating program activities in addressing carbon dynamics on the coastal shelves adjacent to North America, where carbon changes in the terrestrial system greatly influence carbon processes in the coastal ocean.

In the coming years, the integration of terrestrial, oceanic, and atmospheric investigations as a part of NACP and the OCCC program will be a high priority. Assimilation of carbon data into models is developing at scales from regional to global as an important means of incorporating observations into carbon cycle analyses. The goal is to develop increasingly realistic, fully coupled carbon cycle-climate models to provide insight into potential feedbacks between and drivers of these major Earth systems.

Integration of Observations, Research, and Modeling. The primary goal of the Cloud and Land Surface Interaction Campaign (CLASIC) is to improve understanding of the physics of the early stages of cumulus cloud convection as it relates to land surface influences, and to translate this new understanding into improved representations of coupled surface-atmosphere processes in global and regional climate models. Research data from CLASIC will be analyzed in FY 2008 and beyond to address significant uncertainties in climate models, particularly related to their representation of clouds and aerosols. The data from a comprehensive array of measurements from a variety of instrument platforms will be used to characterize the synoptic-scale forcing at the DOE ARM Climate Research Facility's Southern Great Plains site and to undertake modeling studies to establish the most important relationships between land surface conditions and cumulus cloud characteristics. CLASIC is designated as a key focus for interagency coordination in FY 2007 and FY 2008 and beyond by CCSP's Global Water Cycle

IWG. The campaign featured concurrent contributions by NASA, NOAA, and the U.S. Department of Agriculture (USDA) to extend CLASIC's temporal and spatial domain to capture the seasonal time scale and regional processes. The resulting observational framework included ground- and space-based observations, measurements from six airplanes and one helicopter, surface and subsurface hydrologic components, isotopic measurements, CO_2 fluxes, and associated modeling. Planning and operations for CLASIC and the DOE Atmospheric Science Program's Cumulus Humilis Aerosol Processing Study were coordinated. Scientists from CLASIC and the NACP Mid-Continent Intensive Campaign also coordinated measurement and modeling activities. These campaigns represent the cross-cutting activities of three CCSP science elements: the Global Water Cycle, Atmospheric Composition, and Global Carbon Cycle.

Scale and Timing of Climate Forcing. General circulation models, including those with coupled oceans and integrated terrestrial carbon cycles and atmospheric chemistry models, require time-dependent trajectories of GHG emissions and concentrations, chemically active gases, and aerosols to be run in forecast mode. New information from carbon cycle research provides more accurate estimates of the rate of atmospheric CO_2 increase, which translate into a more credible forcing function for climate models. Over the 2008 to 2010 time frame, researchers will build on work reported in CCSP's Synthesis and Assessment Products 2.1a and 2.2 to develop time-dependent trajectories that can be used by climate and atmospheric chemistry models.

Research Plans for Goal 3: 2008 to 2010

CCSP Goal 3: Reduce uncertainty in projections of how the Earth's climate and related systems may change in the future.

OVERVIEW

While a great deal is now known about the mechanisms that affect the response of the climate system to changes in natural and human influences, many questions remain to be addressed and refined. There is still uncertainty regarding precisely how much climate will change overall and especially in specific regions. While the stated focus of the 2003 strategic goal is on reducing uncertainty, it is clear that improving the projections themselves and understanding both the nature and implications of uncertainties are equally important, in order to improve the utility of projections of how Earth's climate and related systems may change in the future. A primary objective of CCSP is to continue to develop information and scientific capacity needed to sharpen both qualitative and quantitative understanding through interconnected observations, data assimilation, and modeling activities. CCSP-supported research will continue to address basic climate system properties and the feedbacks or secondary changes that can either reinforce or dampen the initial and ongoing effects of GHG and aerosol emissions and changes in land use and land cover. The program will also continue to address the potential for future changes in extreme events and uncertainties regarding potential rapid or abrupt changes in climate. CCSP will also continue to build on existing U.S. strengths in climate research and modeling, and to enhance capacity for development of high-end coupled climate and Earth system models.

KEY RESEARCH TOPICS

The 2003 CCSP Strategic Plan highlights key areas in which research progress will help to characterize, understand, and reduce uncertainties in order to provide substantial increases in the utility of projections of how Earth's climate and related systems may change in future. Understanding the scale and timing of climate variability, the couplings among major Earth systems, and the mechanisms and implications of forcings that cause hydrologic changes, drought, and changes in ocean circulation are central to these efforts. Research improvements and outcomes identified in the CCSP Strategic Plan include:

- Improve characterization of the circulation of the atmosphere and oceans and their interactions through fluxes of energy and materials
- Improve understanding of key 'feedbacks' including changes in the amount and distribution of water vapor, extent of ice and the Earth's reflectivity, cloud properties, and biological and ecological systems
- Increase understanding of the conditions that could give rise to events such as rapid changes in ocean circulation due to changes in temperature and salinity gradients

- Accelerate incorporation of improved knowledge of climate processes and feedbacks into climate models to reduce uncertainty in projections of climate sensitivity, changes in climate, and related conditions such as sea level

- Improve national capacity to develop and apply climate models.

In order to better characterize and reduce uncertainty and thereby improve projections of how climate may change in the future, it is essential to improve understanding of basic climate system properties and interactions, including improving characterization of the circulation and interaction of energy in the atmosphere and oceans, and the sources and consequences of uncertainties related to climate feedbacks that can either reinforce or dampen the initial effect of GHGs and aerosols. These feedbacks include changes in ocean circulation, changes in the amount and distribution of water vapor, changes in extent of ice and the Earth's reflectivity, changes in cloud properties, and changes in biological and ecological systems that could significantly change emissions or absorption of GHGs. Although the historical and geologic records clearly show evidence that the climate system can change relatively rapidly, there is considerable uncertainty regarding the potential for changes in extreme events. Rapid, discontinuous changes in climate pose a special problem since they have the potential to cause profound effects on the environment and human well-being since the time available for adaptation would be limited. Important research areas include studies to understand the scale and timing of external climate forcings (e.g., GHG concentration); internal climate forcings (e.g., ENSO events); the dynamics of carbon cycling through Earth's atmosphere, oceans, and land and the potential effects of changes in carbon cycling on climate; and the drivers that may potentially result in rapid, discontinuous changes in ocean thermohaline circulation, drought frequency and distribution, and related hydrologic changes. These research needs are among those addressed by CCSP's 2008 to 2010 research plans. The examples provided below are intended to be illustrative of the breadth of the program's activities under Goal 3, but are not intended to be exhaustive lists of the program's implementation of its strategic directions described above.

ILLUSTRATIVE GOAL 3 PLANS

Atmosphere

Creating a Historical Reanalysis of the Atmosphere of the 20th Century. Recent research has shown the feasibility of using modern data assimilation techniques together with observations of sea-level pressure to produce, for the first time, a global analysis of tropospheric weather patterns at 6-hour temporal resolution that extends over the entire 20th century. Production of this historical reanalysis will be initiated in 2008, with the goal of at least doubling the length of current reanalysis records, which now extend back only until 1948. This historical reanalysis will enable researchers to address such questions as the range of natural variability of high-impact events like floods, droughts, hurricanes, and extratropical cyclones, and how ENSO and other climate modes alter these events. A century-long reanalysis will also help to clarify the origins of climate variations that produced major societal impacts and profoundly influenced policies, including the 1930s 'Dust Bowl' drought and the prolonged cool, very wet period in the western United States early in the 20th century that led to over-allocation of Colorado River water through the 1922 Colorado Compact. This effort will contribute to the development of the Integrated Earth System Analysis capability discussed later in this section.

Land

Integration of Space-Based Observations and Land Surface/ Hydrology Data Assimilation Systems. The Gravity Recovery and Climate Experiment (GRACE) satellite has demonstrated that large-scale changes in the integrated column water content of the combined atmosphere, land surface (including rivers and reservoirs), soil moisture, and groundwater system compare remarkably well with the changes documented by the Global Land Data Assimilation System. In FY 2008 and beyond, further research investigations will explore whether GRACE, A-Train, and other satellite and ground-based data can be assimilated by the Land Information System (LIS) and/or provide integral closure constraints (and updated process parameterizations) to improve the output products from LIS that can potentially be linked to various decision-support tools and systems involved in the management of water resources, among other uses. Such an activity could

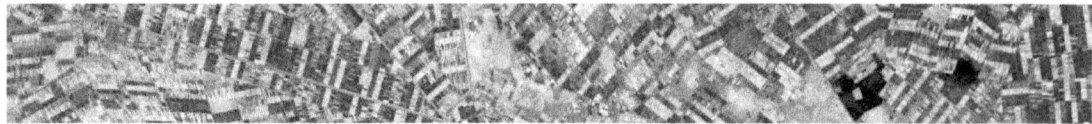

help to identify the initial components of end-to-end capabilities bridging observations, research, modeling, and applications.

Role of Land Surface Processes in North American Hydroclimate. The feedbacks between soil moisture, vegetation, and precipitation will be investigated in observations and models with the goal of helping to understand whether land surface conditions may be a useful predictor in operational climate prediction at seasonal and sub-seasonal time scales. The behavior of snow variations and vegetation cover will be studied in order to improve land surface representations in regional climate models. The hydrologic and climatic effects of crop irrigation are not well quantified and not accurately represented in model initialization. Improvements in our understanding of the role of irrigated croplands in North American hydroclimatic regimes and their representation in models will be pursued.

Advanced Ensemble Multi-Model Hydrological Prediction. Efforts will continue to focus on the calibration and validation of research-mode ensemble (multi-model) forecasting techniques for surface and subsurface hydrological parameters, especially at longer seasonal time scales. The objective is to transfer improved hydrological prediction techniques for operational application at the seasonal to interannual time scale. This activity will expand on the recently developed Advanced Hydrological Prediction Service (AHPS) of NOAA's hydrological forecasting system that includes new model calibration strategies, distributed modeling approaches, ensemble forecasting, data assimilation techniques, enhanced data analysis procedures, flood forecast inundation maps, hydrological routing models, and multi-sensor precipitation estimates. Data will also be ingested from USGS streamflow observations, gridded multi-sensor precipitation and snow water equivalent estimates, and other sources. New approaches for the remote sensing of precipitation, snow, and other inputs will be integrated into the hydrological forecast operation. AHPS is slated to be fully implemented nationwide in 2013. In parallel, CCSP researchers plan to participate in the further development of the international Hydrological Ensemble Prediction Experiment, which will bring the international hydrological community together with the meteorological community and demonstrate how to produce reliable hydrological ensemble forecasts that can be used with confidence by emergency management and water resources sectors to make decisions that have important consequences for the economy, and for public health and safety.

Land-Oceans-Atmosphere Integration (Climate System)

Constructing a Satellite-Era Reanalysis of the Coupled Ocean-Atmosphere System. A national capacity for Integrated Earth System Analysis (IESA) is being developed that extends beyond current attempts to map individual components of the Earth system separately (see the IESA description in the Interagency Implementation Plans at the end of this section). Achieving this capability requires parallel advancements in coupled Earth system modeling, and considerable progress is being made in this latter arena, particularly with the adoption by the research and operational forecasting communities of a common Earth system modeling framework. A coupled ocean-atmosphere model is currently in operational use, and beginning in 2008 this model will serve as the basis for the first attempt to create a reanalysis of the coupled ocean-atmosphere system dating back to the start of the satellite era (1979) through 2007. Development of a coupled ocean-atmosphere analysis capability will also support intensified efforts to improve the monitoring and understanding of changes in the ocean thermohaline circulation.

Advanced Carbon Models. Research will continue to develop carbon cycle and coupled carbon-climate models that are more comprehensive in their treatment of significant carbon dynamics and drivers, including those involving or stemming from human activities. These advanced models will incorporate multiple, interacting factors, address time scales of decades to centuries, and integrate across spatial scales. Carbon data assimilation at both regional and continental scales will continue to integrate multiple data streams including fluxes by eddy-covariance methods, atmospheric CO_2 and methane concentrations using tower-based instruments, and aircraft and satellite remote-sensing observations. An important aspect of these integrated investigations is the potential to identify drivers within the carbon cycle and climate system, and especially in those systems where sensitivities of carbon processes and stocks to climate change are high (e.g., Arctic and boreal systems). Modeling research focusing on the Southern Ocean and Antarctica will capitalize on existing and impending remote and *in situ* observations and will include synthesized data sets, existing models, and data assimilation techniques to advance the ability to quantify southern high-latitude sensitivities and variability. Significant advances are expected in regional and global carbon cycle modeling.

Drought in Coupled Models Project. A new, multi-agency activity will support research into the physical and dynamical mechanisms of drought and the mechanisms through which drought may change as climate changes. Relevant issues include the role of the seasonal cycle in drought, the impacts of drought on water supplies, and the distinction between drought as a transient phenomena and drying produced by long-term changes in a region's water balance. A broad range of model simulations will be analyzed and evaluated in this effort, including multi-model simulations of 20th-century climate, model projections of future climate, paleoclimate simulations of the last glacial maximum, and seasonal model prediction data sets. The objective is to increase community-wide diagnostic research into the physical mechanisms of drought and to evaluate drought simulations by current models. This effort will lead to more robust evaluations of model projections of drought risk and severity, and to a better quantification of the uncertainty in such projections.

Climate Variability and Predictability Modeling. U.S. CLIVAR is exploring a new approach for bringing together observers, theorists, and high-end modelers to improve key model deficiencies (USCLIVAR, 2002; Bretherton et al., 2004). This approach is attempting to significantly reduce the time lags that often exist between observation of key climate processes and integration of these processes into more comprehensive Earth system models. Several high-end Earth system modeling efforts in the United States, which involve many different, independent research teams, are using these types of new collaborative approaches and tools to evaluate, improve, and integrate model components.

Research Plans for Goal 4: 2008 to 2010

CCSP Goal 4: Understand the sensitivity and adaptability of different natural and managed ecosystems and human systems to climate and related global changes.

OVERVIEW

Seasonal to annual variability in climate has been connected to impacts on ecosystems and many aspects of human life. Longer time scale natural climate cycles and human-induced changes in climate have additional effects.

Improving the ability to assess potential implications of variations and future changes in climate and environmental conditions on terrestrial, freshwater, and marine ecosystems and physical systems (including ocean chemistry) and human systems (especially human health, human welfare, and communities) could enable governments, businesses, and communities to mitigate damages and to seize opportunities by adapting infrastructure, activities, and plans. CCSP research will continue to examine these multiple interacting changes and effects (e.g., changes in climate and climate variability coupled with the CO_2 'fertilization effect,' deposition of nitrogen and other nutrients, changes in landscapes that affect water resources and habitats, changes in frequency of fires or pests, or how changes in the climate system could impact various sectors and regions in order to identify populations and regions that are likely to be particularly vulnerable to specific impacts) in order to improve knowledge of sensitivity and adaptability of systems to climate variability and change. CCSP research will also improve methods to integrate understanding of potential ecological effects of different atmospheric concentrations of GHGs, and to develop methods for aggregating and comparing potential impacts across different sectors and settings.

KEY RESEARCH TOPICS

The 2003 CCSP Strategic Plan identifies the key research questions that must be answered in order to better understand the sensitivity and adaptability of managed and unmanaged ecosystems and human systems to climate and related global changes. Important components of this goal include the understanding that the potential effects of climate variability and change on ecosystems and human activities are a composite derived from their sensitivity and adaptability together with complex interactions among physical, ecological, economic, and social systems and conditions. The development of a better understanding of ecosystem and human system responses requires in turn that we attain a better understanding of the contributing factors that influence them, and are influenced by them. Needed research outcomes identified in the CCSP Strategic Plan include:

- Improve knowledge of the sensitivity of ecosystems and economic sectors to global climate variability and change

- Identify and provide scientific inputs for evaluating adaptation options, in cooperation with mission-oriented agencies and other resource managers

- Improve understanding of how changes in ecosystems (including managed ecosystems such as croplands) and human infrastructure interact over long periods of time.

A sensitive system is one that responds quickly and perhaps dramatically to change. Fundamental to the understanding of the potential effects of climate variability and change on ecosystems and human activities is the understanding that these effects will be determined both by the sensitivity and adaptability of ecosystems and human activities plus multiple, cumulative, complex interactions among the physical, ecological, economic, and social conditions that provide the framework in which these ecosystems and human activities reside. Ecosystems and human systems alike face potential multiple stressors as climate changes, and system resiliency plays a strong role in determining system response. Among the highest priority research required to improve understanding of sensitivity, resiliency, and adaptability is better understanding of the ways in which terrestrial ecosystems control and respond to changes in soil moisture, temperature, and vegetation to exert an influence on precipitation and surface hydrology; the effects of land use and land-use change on sensitive systems; and the overall effects on sensitive systems of changes in precipitation that may result from climate variability and change.

Further priorities include a better understanding of the potential impact of ocean acidification on coral reefs, marine food webs, and ocean carbon balance, as well as the implications of potentially decreased sea ice extent for polar ecosystems and communities, and the impact of

potential freshwater injection from high-latitude regions into the polar oceans on marine ecosystems. Additional topics requiring study include a better understanding of the potential consequences of various carbon management strategies, and an understanding of potential climate change forcings, feedbacks, and effects. CCSP will emphasize improving the understanding of these topics at a regional scale. Studies that address these subjects are included in the illustrative 2008 to 2010 CCSP research plans outlined below. (Note also that additional plans that cut across elements of Goals 4 and 5 can be found in the Goal 5 plans later in this section.) These examples are intended to be indicative of the breadth of the program's activities under Goal 4, but are not intended to be exhaustive lists of the program's implementation of its strategic directions described above.

ILLUSTRATIVE GOAL 4 PLANS

Oceans

Ocean Acidification: Changing Oceans, Changing Ecosystems. Rising atmospheric CO_2 levels are altering ocean chemistry and threatening marine biodiversity. Decreasing oceanic pH resulting from increasing atmospheric CO_2 reduces the abilities of calcifying organisms, such as corals and crustaceans, to form skeletons and shells. It is increasingly urgent to understand and predict their likely responses to increased acidity (reduced pH). Upcoming research will include studies on evolutionary change in organisms in response to changing chemistry, compensatory shifts in species within functional groups and the role of biodiversity in facilitating such shifts, effects of lower pH on coral calcification, functional genomics studies of pH effects on molecular regulation of calcification, and development of models to predict effects of multiple environmental changes on organism-, population-, and ecosystem-level adaptation.

Climate Impacts on Marine Ecosystems. Projects will be implemented to understand the responses of the physical environment and ecosystems to projected climate change scenarios and to develop predictions of the ecosystem impacts of these changes. These projects will collect observations, conduct research, and synthesize results to increase the understanding of regional climate impacts on marine ecosystems. This work will be conducted in conjunction with the development and refinement of biophysical indicators and models to provide living marine resource managers the knowledge and predictive tools necessary to adapt to the consequences of climate change for ecosystems.

Land

Effects of Changes in Precipitation on Southwestern Ecosystems. Climate models indicate that precipitation and soil moisture are likely to change in the southwestern United States during this century. To reduce scientific uncertainty about the potential effects of such changes on the structure and functioning of terrestrial ecosystems throughout the region, field experiments involving *in situ* manipulation of precipitation and soil moisture will be conducted in southwestern pinyon-juniper woodlands, coast sage, grassland, chaparral, and oak-pine forest ecosystems. Measurements will elucidate potential effects of altered precipitation on primary production processes, species diversity, decomposition of soil organic matter and related biogeochemistry, and aspects of ecological feedbacks to the physical climate system.

Integrated Impacts on and Adaptation to Climate Change of Terrestrial Ecosystems, Water Resources, and Agriculture. Terrestrial ecosystems, water resources, and agriculture represent important systems and pathways through which climate change could be experienced. Integrated models of drivers and systems response have generally lagged progress on the discrete research topics and associated models. Beginning in FY 2008, foundations will be laid for incorporating these features into integrated models in a way that reflects the state-of-the-art in relevant disciplinary research. Scoping meetings will be conducted to explore the state-of-the-art, identify methods for incorporation, and deliver a research plan. In addition, first steps will be taken to implement that plan, including a preliminary evaluation of climate impacts on water resources with implications for agricultural impacts and adaptation.

Human Health

Climate Change and Human Health. Exposure to allergens results in allergenic illnesses in approximately 20% of the U.S. population. Climate change, including increased atmospheric CO_2 concentrations, could have significant impacts on the production, distribution, dispersion, and allergenicity of aeroallergens and the growth and distribution of organisms that produce them (i.e., weeds, grasses, trees, and fungi). Shifts in aeroallergen production and, subsequently, human exposures, may result in changes in the prevalence and severity of symptoms in individuals with allergenic illnesses. CCSP agencies plan to investigate this potential health effect through the award of competitive multiyear grants as part of an interagency program on climate change and health.

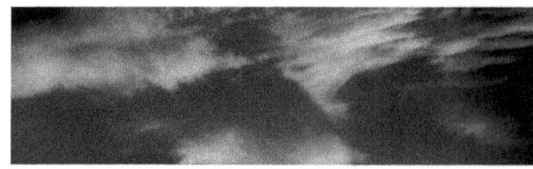

Research Plans for Goal 5: 2008 to 2010

CCSP Goal 5: Explore the uses and identify the limits of evolving knowledge to manage risks and opportunities related to climate variability and change.

OVERVIEW

In recent years, the scientific and technical community has begun to develop a variety of products to support management of risks and opportunities related to climate variability and change, but much remains to be done in this area. CCSP will foster additional studies and encourage evaluation and learning from these experiences in order to develop and improve decision-support processes and products that use knowledge to the best effect, while communicating levels of uncertainty appropriately. Working in partnership with stakeholders and end users of this information, CCSP will develop resources (e.g., observations, databases, data and model products, scenarios, visualization products, scientific syntheses, assessments, tools, and approaches to engage ongoing consultative mechanisms) to support scientifically informed policymaking, planning, risk reduction, and adaptive management.

KEY RESEARCH TOPICS

In order to be useful to policymakers and managers, scientific information from multiple sources must be brought together in the form of comparative analyses, scenarios, and tools that evaluate the impacts of potential change and the consequences of potential short- and long-term policies and decisions. In all cases, the science that underpins the tools must be robust and uncertainties must be well characterized. The 2003 CCSP Strategic Plan outlines three key approaches that must be met in order to provide a sound basis for decisionmaking with regard to the effects of potential climate change:

- Support informed public discussion of issues of particular importance to U.S. decisions by conducting research and providing scientific synthesis and assessment reports

- Support adaptive management and planning for resources and physical infrastructure sensitive to climate variability and change, and build new partnerships with public and private sector entities that can benefit both research and decisionmaking

- Support policymaking by conducting comparative analyses and evaluations of the socioeconomic and environmental consequences of response options.

Decision support provides the connection between the science of understanding climate change and its impacts, and the use of that understanding to inform the process by which decisions are made to adapt to and mitigate climate effects and to reduce risk and increase resiliency in Earth systems and human communities. The three approaches outlined above – assessments, adaptive management, and policy-relevant analyses – provide the bridge to put science in the hands of managers and policymakers. The need for products that integrate and assess climate-related changes and that provide an evaluation of the socioeconomic and environmental effects and consequences of potential climate change is a major driving force behind three of CCSP's key activities: 1) participation in and support of the IPCC Fourth Assessment Report; 2) production of the 2006 to 2008 synthesis and assessment products, which include analyses of impacts and consequences; and 3) CCSP involvement in adaptive management.

Meeting the challenges that face the policymaking and land and resource management and energy production communities involves ongoing, real-time learning and knowledge creation, both in a substantive sense and in terms of the process itself, and requires that stakeholder communities be involved throughout the process rather than at the end. Adaptive management focuses on learning and adapting through partnerships of managers, scientists, and other stakeholders who learn together how to maintain sustainable ecosystems. Decision support at scales relevant to land and resource managers requires the development and implementation of regionally resolved models with accuracy sufficient to support sound decisionmaking. The end goal of these decision-support activities is to engage user communities and provide managers and policymakers with the knowledge required to understand ecological and human systems and the uncertainties inherent in changing conditions, to manage them under those changing conditions, and to make decisions that maximize flexibility to reassess and adjust their management approaches as needed.

Crucial areas in which decision support is needed include the understanding of effects of changing climate in coastal zones, where sea-level rise and changes in storm intensity and frequency may occur; in urban areas where effects on stream drainage and potential flood impacts may be felt; in the management of aerosols and other emissions related to air quality, ozone depletion, and radiative transfer from the atmosphere to the land; in understanding the effects and feedbacks of potential climate change and variability (e.g., drought and/or storminess) on the land surface, on water quality and availability, and in understanding the effects of land-use changes related to the consequences of carbon sequestration and bioenergy development on ecological goods and services. These are among the topics addressed by CCSP's 2008 to 2010 research plans as illustrated below. These examples are intended to be indicative of the breadth of the program's activities under Goal 5, but are not intended to be exhaustive lists of the program's implementation of its strategic directions described above.

ILLUSTRATIVE GOAL 5 PLANS

Oceans and Coastal Regions

Decision Assessment in the Gulf Coast and Chesapeake Bay Regions. Several pilot studies in the Gulf Coast region and the Chesapeake Bay were initiated to test different approaches to assessing the flow and use of climate change science information in decisionmaking, the factors and institutions that affect its use, and the types and characteristics of decisions most sensitive to climate change and most in need of additional reevaluation and research in light of projected changes. CCSP researchers plan to evaluate the results of these pilot studies to determine the applicability of a decision assessment approach to decisions related to water quality, aquatic ecosystems, and air quality.

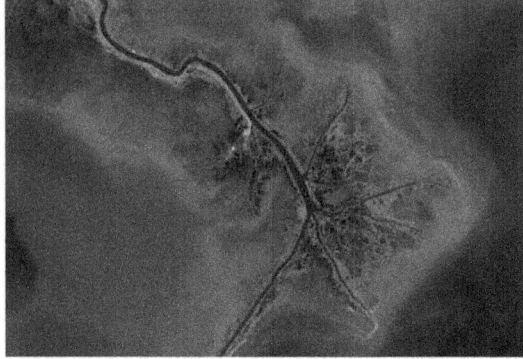

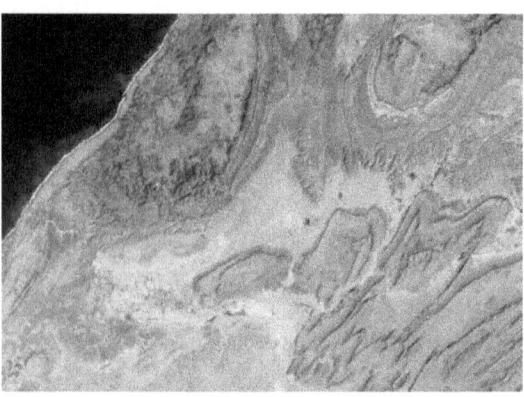

Land

Continued Development of Tools for the Assimilation of Remote-Sensing Data into Water Quality and Sediment Transport and Erosion Models. Agricultural research activities in the area of land data assimilation systems and model analysis are focused on the efficient integration of ground-based and remote-sensing data into critical resource and conservation practice assessment models. Existing agency research projects are aimed at the sequential assimilation of surface soil moisture retrievals and vegetation indices from microwave and visible remote sensors to constrain crop growth and root-zone water balance models. In FY 2008 and beyond, this work will expand with an emphasis that includes the assimilation of remote-sensing data into distributed water quality and sediment transport and erosion models. Particular attention will be paid to development of data assimilation and modeling capabilities to quantify benefits arising from the adoption of conservation practices within agricultural watersheds.

Development of Land-Use Change Models. Developing future land-use and land-cover change (LULCC) scenarios and understanding drivers and feedbacks are necessary to assess impacts on the environment. A tool to generate future LULCC scenarios consistent with IPCC emissions scenarios is being developed, and will be used to assess the environmental impacts of climate and land-use change regionally and nationally. This will improve projections of climate and global change and contribute to understanding of possible management risks and opportunities related to climate change. It also contributes directly to understanding the feedbacks between climate change, conservation policies, and land-use and land-cover decisions.

New Regional Integrated Sciences and Assessments Effort Focused on Drought. NOAA's Regional Integrated Sciences and Assessments (RISA) program, established in the mid-1990s, supports research that addresses complex climate-sensitive issues of concern to decisionmakers and policy planners at a regional level. RISA research team members are primarily based at universities, although some of the team members are based at government research facilities, nonprofit organizations, or private sector entities. Traditionally the research has focused on the fisheries, water, wildfire, and agriculture sectors, as well as climate-sensitive public health and coastal restoration issues. Under the auspices of the Coping with Drought through Research and Regional Partnerships effort described in the FY 2007 edition of *Our Changing Planet*, a new activity will be initiated in a region not currently covered by the RISA program. This new RISA activity will have drought impacts research and stakeholder work as a central theme and will provide an avenue within the chosen region for interagency work focused on climate impacts.

Development of Modeling Tools to Support Water and Watershed Management. Climate change presents a range of risks and opportunities to water managers. To minimize risk and take advantage of opportunities, tools are necessary to promote adaptive and forward-looking environmental management by decisionmakers at all levels. In 2007, a new climate assessment capability was developed within the Better Assessment Science Integrating Point and Non-point Sources (BASINS) watershed modeling system. The new tool facilitates assessment of the influence of climate variability and change on water quantity and quality and provides a capacity to evaluate adaptation strategies that increase the resilience of water systems to changes in climate. A case study using the new BASINS system is underway in 2008 and beyond to assess the sensitivity of hydrologic and water quality endpoints to climate change in the Monocacy River watershed, a tributary of the Potomac River and Chesapeake Bay. An on-line decision-support capability with the USDA Agricultural Research Service Water Erosion Prediction Project (WEPP) soil erosion model also is under development. New climate change assessment capabilities within WEPP will enable land managers to develop best management practices to lessen the impacts of climate variability and change on sediment loading from agricultural land to streams. As this project goes forward, the need for developing similar climate assessment capabilities for models applicable to urban drainage and design will be determined.

Integrated Evaluation of Climate Change, Mitigation, Bioenergy, and Land Use. Biofuels represent a potentially important source of energy that, depending on how they are produced, could reduce CO_2 emissions by replacing

fossil fuels. However, greatly expanded use of biofuels would put pressure on food and forestry prices and could lead to changes in land use and consequent release of carbon from soils and vegetation. At the same time, changes in climate, CO_2 levels, and concentrations of other pollutants such as ozone could affect the productivity of crops (including bioenergy feedstocks), pasture, and forest land. In 2008 and beyond, researchers will complete linkage of a multi-sector, multi-region general equilibrium model of the world economy with a terrestrial ecosystem model that simulates biogeochemical processes of land systems at a 0.5° latitude-longitude grid level. The linkage will allow examination of the effects on GHG cycles of disturbances associated with the conversion of unmanaged forest and grassland to crops, pasture, or bioenergy feedstock production. Also, because it will be fully integrated with economic projections, the linked system will provide the ability to evaluate the feedbacks of changing environmental conditions to agricultural productivity, the resultant effects on the global and regional economy, consequent impacts on land use, and the further effects of land use change on biogeochemical cycles and feedbacks to atmospheric composition and climate.

Human Health

Valley Fever Public Health Decision-Support System. Valley Fever (*coccidioidomycosis*) is a disease endemic to arid regions in the Western Hemisphere, and is caused by soil-dwelling fungi. Arizona is currently experiencing an epidemic, with almost 4,000 cases in 2004, greatly exceeding other climate-influenced diseases such Hantavirus or West Nile Virus. The fungus responds to changes in climate conditions, such as precipitation and atmospheric dust. Climate models and satellite-derived spatial data on soil moisture and land-cover disturbance are being used to evaluate seasonal associations with Valley Fever incidence. Working in partnership with the Arizona Department of Health Services, CCSP scientists are developing a decision-support tool to make seasonal orecasts of disease incidence by geographic area and display spatial relationships with environmental conditions.

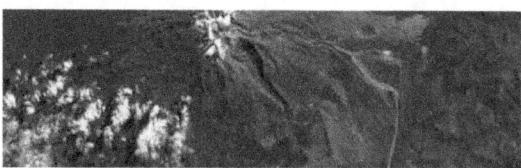

Interagency Implementation Plans

As discussed above, significant research questions remain to be articulated and answered across all of CCSP's goals. While many important research questions are specific to a single CCSP strategic goal, there are others that inherently speak to more than one of the goals.

Each of the 13 CCSP-participating agencies has its own priorities, plans, and relevant activities that make invaluable contributions to CCSP, and which contribute a large portion of CCSP's progress toward its strategic goals. A summary of individual agency priorities can be found in CCSP's annual report to Congress. However, some research questions are so large and multivariate that no single agency can effectively undertake to answer them; rather, they require the cooperation of multiple agencies and agency programs. One mechanism by which CCSP addresses these integrated endeavors is through the development of near-term (i.e., 1-3 year) interagency implementation plans. An example of a near-term interagency implementation plan that CCSP has identified as needing intensive effort is a focus on understanding carbon cycling and climate change in high-latitude regions, since these regions are among the most rapidly changing areas of the planet. Another example is the development of an integrated Earth system analysis capability to focus on creation of a high-quality record of the state of the atmosphere and ocean since 1979, information that is needed in order to improve the assimilation of land cover and dynamic sea ice modeling into carbon and nutrient cycling and other crucial areas.

While these implementation plans are only a part of the overall program, they are vital mechanisms through which CCSP integrates agency activities to create knowledge and products that are greater than the sum of the individual agency efforts. Because of their multivariate nature, they do not fit easily within the preceding goal-by-goal articulations of research plans; rather, because of their multidisciplinary, integrative nature, each implementation priority contributes to multiple goals. The following are selected examples of implementation priorities for the coming years that are inherently interagency and multi-goal. For more specific details on these implementation priorities, see CCSP's annual report to Congress (CCSP, 2007b, 2008).

Enhanced Carbon Cycle Research in High-Latitude Systems. The global carbon cycle has been one of the seven interdisciplinary science focus areas for the U.S. Global Change Research Program (USGCRP) for many years. Accomplishments include completion of CCSP Synthesis and Assessment Product 2.2, *The First State of the Carbon Cycle Report,*[a] as well as improved availability of CO_2 measurements, plus advances in coupled carbon-climate modeling and assimilation. Recognition that high-latitude systems are increasingly important sources of atmospheric carbon as regional warming occurs makes it a high priority to improve understanding of the carbon dynamics in high-latitude systems, and the factors that may lead to changes in those dynamics. These are crucial elements of global carbon modeling and a priority for understanding the linkages and feedbacks between carbon, ecosystems and land cover, hydrology, and climate variability and change.

Quantification of Climate Forcing and Feedbacks by Aerosols, Non-CO_2 Greenhouse Gases, Water Vapor, and Clouds. The need to quantify and understand the impacts of radiative forcing on climate has long been important to USGCRP. Advances have been made in understanding of climate influences of aerosols, reactive gas emissions and transformations, and ozone in both the troposphere and stratosphere, and these continue to be important. The next level of complexity adds the importance of water vapor in the upper troposphere and lower stratosphere as a key component of the atmospheric system. Additional work is needed to further improve quantification of the climate forcing associated with aerosols, clouds (including cirrus), the spatially varying shorter lived trace gases, and upper tropospheric and lower stratospheric ozone. Recent analysis, including that associated with the Fourth Assessment Report of the IPCC, has emphasized this need, and a number of scientific advances and improvements in observation and modeling capability make the timing appropriate for an enhanced focus on this topic.

Development of an Integrated Earth System Analysis (IESA) Capability: A Focus toward Creating a High-Quality Record of the State of the Atmosphere and Ocean Since 1979. Just as the public and decisionmakers can today easily access weather maps (i.e., 'analyses' of the atmosphere) to support a wide range of applications, tomorrow's decisionmakers need tools to visualize the evolving state of the climate system over the entire planet, including its oceans, land surface, and vegetation. By combining global observations of the atmosphere, ocean, land, biosphere, and ice-covered

---

[a] See <www.climatescience.gov/Library/sap/sap2-2/final-report/default.htm>.

areas with models that dynamically couple these components of the Earth system, it will be possible to produce internally consistent maps (i.e., 'analyses') of the state of the planet. IESA is in the beginning stages of developing this capability, and the initial data assimilation components of the project are discussed earlier in this section of the Revised Research Plan, under Goal 3 plans. Reanalysis activities in FY 2008 will include an improved treatment of the hydrological cycle and efforts to extend atmospheric reanalyses to span the entire 20th century. Future advances in producing integrated Earth system analyses will require progress in ongoing efforts to construct models that properly simulate the interactions among the physical and biogeochemical processes in the climate system. Time series of such analyses will allow researchers to better explain observed changes in the climate system and will allow decisionmakers to develop more informed options to address future changes.

Development of an End-to-End Hydrologic Projection and Application Capability. The need to provide information to water resource managers and other decisionmakers on issues related to how climate affects water availability, drought, and water quality has long been a component of CCSP activities, and the global water cycle is one of CCSP's identified research elements. An end-to-end system to provide information to water resource managers and other decisionmakers on issues related to how climate affects water availability, drought, and water quality requires integration and improvement of existing research and monitoring capabilities to reduce uncertainties in hydrological/climate predictions. Assembling the building blocks for the development of an end-to-end global water cycle infrastructure and the development of an observations-based Generalized Hydrological (water, energy, biogeochemical) Modeling/Prediction Framework

will help to reduce uncertainties and improve hydrologic predictions, leading to improved decision-support information and resources.

Assessing Abrupt Change in a Warming Climate: Toward Development of an Abrupt Change Early Warning System. Changes in the climate system are considered 'abrupt' if they occur more rapidly than the time needed by society and ecosystems to adapt to them (NRC, 2002). Paleoclimate research indicates that major shifts in regional and global climate have occurred on time scales as short as decades, severely affecting rainfall patterns, droughts, ecosystems, and human civilizations (IPCC, 2007). Possible impacts range from accelerated melting of ice sheets and associated sea-level rise, to severe and sustained droughts, to systematic changes in weather patterns over broad regions that may result from changes in ocean circulation. Assessing the potential for future abrupt changes and implementing the capability to diagnose and predict their occurrence will require concerted efforts to improve Earth system analysis, decadal forecasting capabilities, reconstructions of past abrupt climate change, and understanding of societal impacts. CCSP has a research element aimed specifically at climate variability and change, which has fostered considerable progress in understanding of past abrupt climate events and the potential causes for rapid changes, and Synthesis and Assessment Product 3.4, *Abrupt Climate Change*, which includes discussion of potential risks associated with abrupt climate change, is scheduled for publication in 2008 (see Appendix 1). Activities during FY 2008 to FY 2010 will emphasize model experiments designed to test potential mechanisms for abrupt change, and paleoclimate research on patterns, causes, and impacts of past abrupt climate events. Both activities will help set priorities for enhanced monitoring with the goal of developing an abrupt change

early warning system. These efforts are integrated with the National Science and Technology Council's Ocean Research Priorities Plan near-term priority on the meridional overturning circulation of the Atlantic Ocean (NSTC, 2007).

Ecological Forecasting. Ecological forecasting brings together modeling with observations and results from experiments and process studies to predict the impacts of natural and anthropogenic environmental changes on life-sustaining ecosystems. Many CCSP agencies are engaged in activities that include components of an ecological forecasting capability to address emerging questions.

Progress has been made in such areas as documenting changes occurring in boreal forests. This has set the stage for reducing scientific uncertainty about possible future changes in primary production, biogeochemistry, and biodiversity, and in findings that show that global oceanic phytoplankton productivity responds to changes in upper ocean temperature and stratification. Work in the coming years will build upon earlier investigations to expand the development of models linking geophysical and ecological phenomena, to better characterize the uncertainty associated with linked models, and thus to provide more reliable ecological forecasts. The result will be an enhanced understanding of ecological response to changing climate as well as improved natural resource management and decisionmaking.

VII Conclusion

VII. Conclusion

This Revised Research Plan provides – for both programmatic and strategic science goals – a brief summary of CCSP progress and a snapshot of CCSP plans with illustrative examples of research, products, and outreach as envisioned for the coming 3 years and beyond. Taken together, these activities will build upon the work already done to provide a solid foundation for the Nation's needs in understanding, adapting to, and mitigating the effects of climate change.

For each of CCSP's strategic goals, the Revised Research Plan provides a series of specific examples of research priorities and plans. **The illustrative examples in the Revised Research Plan are not intended to provide an exhaustive list of every CCSP research project, but rather they provide an indication of the breadth and depth of the program's activities.** In addition to the illustrative research examples identified, it is fully expected that other important research topics, yet to be determined, will emerge from future scientific progress, events, and societal needs.

As identified in the Revised Research Plan, key components of CCSP's activities over the next 3 years include the following:

- CCSP will continue to provide the basic physical science required to understand Earth's past and present climate, including its natural variability, and to improve understanding of the causes of and uncertainties in observed variability and change at global, continental, regional, and local scales. CCSP remains committed to basic, ongoing research to understand climate processes and the forcing factors that cause changes in climate and related systems.

- CCSP will increasingly address emerging needs for research to more fully understand the impacts of climate change on unmanaged and managed ecosystems, human health and infrastructure, and economic and other human systems.

- CCSP will continue to generate science in support decisionmaking related to the management of risks and opportunities of climate variability and change, including adaptive management and mitigation efforts, with an increased emphasis on generating scientific results at regional and local scales.

- CCSP will place greater emphasis on communicating with users and stakeholders (e.g., state and local governments, academia, industry, public utilities,

and nongovernmental organizations), both to gain the benefit of their experience, perspectives, and input and to ensure that the results of CCSP research, monitoring data, and assessments are widely and easily available and accessible to potential users of this information.

These four points distill the key similarities and differences between the 2003 CCSP Strategic Plan and the way forward that is identified and illustrated in the Revised Research Plan.

This document represents one step in CCSP's ongoing planning process, and reflects input from CCSP's member agencies, the NRC through its reports, the scientific community via CCSP IWGs and Science Steering Groups, and comments provided by the public. As this document is being published, longer term strategic planning efforts are already underway. This multi-year process will culminate in the production of CCSP's next Strategic Plan.

This Revised Research Plan is a living document, part of the longer range self-study to articulate progress and to plan for the future. CCSP research necessarily evolves to take into consideration emerging societal and scientific needs; the changes and shifts in emphasis in major scientific questions that have resulted from advances in knowledge and other accomplishments; the most urgent research needs that have emerged; and the expected outcomes, products, and impacts for maximum societal benefits. The examples provided herein demonstrate the work it will undertake and the direction that CCSP will evolve in the near future, toward increased engagement with stakeholders and increased attention to relevance of scientific results to decisionmaking and policymaking.

CCSP's role in meeting the challenges of global change is vital. CCSP adds value to Federal agency efforts in climate change research and related activities by providing a structure and coordination mechanism that leverages individual agency efforts through increased cooperation, collaboration, and the joint development of research priorities. The CCSP framework allows Federal agencies engaged in global change research to do more than individual agencies could do separately, and thus to more effectively address the Nation's global change science needs and coordinate and communicate activities with those of its domestic and international research partners and stakeholders. In contributing to this framework, the Revised Research Plan is consistent with the CCSP guiding vision of "a Nation and the global community empowered with the science-based knowledge to manage the risks and opportunities of change in the climate and related environmental systems."

VIII

Information Regarding the Preparation of the Revised Research Plan

VIII. Information Regarding the Preparation of the Revised Research Plan

External Expert Reviewers

The helpful comments of the following external expert reviewers are gratefully acknowledged:

Roberta Balstad	CIESIN, Columbia University
David Behar	Water Utility Climate Alliance
Kris Ebi	Consultant
Dennis Hartmann	University of Washington
Bev Law	Oregon State University
Margaret Leinen	Climos
John Novak	The Electric Power Research Institute
Dennis Ojima	The Heinz Center
Michael Prather	University of California - Irvine
David Rind	NASA Goddard Institute for Space Studies
John Walsh	University of Alaska - Fairbanks

References Cited

ACIA, 2005: *Arctic Climate Impact Assessment.* Cambridge University Press, 1042 pp.

Angert, A., S. Biraud, C. Bonfils, C.C. Henning, W. Buermann, J. Pinzon, C.J. Tucker, and I.Y. Fung, 2005: Drier summers cancel out the CO_2 uptake enhancement induced by warmer springs. *Proceedings of the National Academy of Sciences*, **102**, 10823-10827.

Barnett, T.P., D.W. Pierce, K.M. AchutaRao, P.J. Gleckler, B.D. Santer, J.M. Gregory, and W.M. Washington, 2005: Penetration of human-induced warming into the world's oceans. *Science*, **309**, 284-287.

Bates, N.R. and A.J. Peters, 2007: The contribution of acid deposition to ocean acidification in the subtropical North Atlantic Ocean. *Marine Chemistry*, **107**, 547-558.

Birdsey, R.A., 2006: Carbon accounting rules and guidelines for the United States forest sector. *Journal of Environmental Quality*, **35**, 1518-1524.

Boote, K.J., L.H. Allen Jr., P.V. Prasad, J.T. Baker, R.W. Gesch, A.M. Snyder, D. Pan, and J.M. Thomas, 2005: Elevated temperature and CO_2 impacts on pollination, reproductive growth, and yield of several globally important crops. *Journal of Agricultural Meteorology*, **60**, 469-474.

Bretherton, C.S., R. Ferrari, and S. Legg, 2004: Climate Process Teams: a new approach to improving climate models. *U.S. CLIVAR Variations Newsletter*, **2(1)**, 1-5.

Cahoon, D.R., P.F. Hensel, T. Spencer, D.J. Reed, K.L. McKee, and N. Saintilan, 2006: Coastal wetland vulnerability to relative sea-level rise: wetland elevation trends and process controls. In: *Wetlands as a Natural Resource, Volume 1: Wetlands and Natural Resource Management* [Verhoeven, J., D. Whigham, R. Bobbink, and B. Beltman (eds.)]. Springer Ecological Studies Series Vol. 190, Springer, pp. 271-342.

CCSP, 2003: *Strategic Plan for the U.S. Climate Change Science Program.* U.S. Climate Change Science Program, Washington, DC.

CCSP, 2005: *Climate Science in Support of Decision Making.* U.S. Climate Change Science Program, Washington, DC.

CCSP, 2006: *Temperature Trends in the Lower Atmosphere: Steps for Understanding and Reconciling Differences.* A Report by the U.S. Climate Change Science Program and the Subcommittee on Global Change Research [Karl, T.R., S. Hassol, C.D. Miller, and W.L. Murray (eds.)]. National Oceanic and Atmospheric Administration, National Climatic Data Center, Asheville, NC, 164 pp.

CCSP, 2007a: *North American Carbon Budget and Implications for the Global Carbon Cycle.* Synthesis and Assessment Product 2.2 – A Report by the U.S. Climate Change Science Program and the Subcommittee on Global Change Research [Dilling, L., A. King, D. Fairman, R. Houghton, G. Marland, A. Rose, T. Wilbanks, and G. Zimmerman (eds.)]. National Oceanic and Atmospheric Administration, National Climatic Data Center, Asheville, NC, 242 pp.

CCSP, 2007b: *Our Changing Planet.* U.S. Climate Change Science Program, Washington, DC.

CCSP, 2008: *Our Changing Planet.* U.S. Climate Change Science Program, Washington, DC.

Chandler, M.A., H.J. Dowsett, and A.M. Haywood, 2008: The PRISM Model-Data Cooperative: Mid-Pliocene data-model comparisons. *PAGES News*, **16**, 24-25.

Collins, W.D., C.M. Bitz, M.L. Blackmon, G.B. Bonan, C.S. Bretherton, J.A. Carton, P. Chang, S.C. Doney, J.J. Hack, T.B. Henderson, J.T. Kiehl, W.G. Large, D.S. McKenna, B.D. Santer, and R.D. Smith, 2006: The Community Climate System Model: CCSM3. *Journal of Climate*, **19**, 2122-2143.

Cook, E.R., C.A. Woodhouse, C.M. Eakin, D.M. Meko, and D.W. Stahle, 2004: Long-term aridity changes in the Western United States. *Science*, **306**, 1015-1018.

Cronin, T.M., H.J. Dowsett, G.S. Dwyer, P.A. Baker, and M.A. Chandler, 2005: Mid-Pliocene deep sea bottom water temperatures based on ostracode Mg/Ca ratios. *Marine Micropaleontology*, **54**, 249-261.

Curry, R. and C. Mauritzen, 2005: Dilution of the northern North Atlantic in recent decades. *Science*, **308**, 1772-1774.

Daughtry, C.S.T., P.C. Doraiswamy, E.R. Hunt, A.J. Stern, J.E. McMurtrey, and J.H. Prueger, 2006: Remote sensing of crop residue cover and soil tillage intensity. *Soil and Tillage Research*, **91**, 101-108.

Delworth, T.L., A.J. Broccoli, A. Rosati, R.J. Stouffer, V. Balaji, J.A. Beesley, W.F. Cooke, K.W. Dixon, J. Dunne, K.A. Dunne, J.W. Durachta, K.L. Findell, P. Ginoux, A. Gnanadesikan, C.T. Gordon, S.M. Griffies, R. Gudgel, M.J. Harrison, I.M. Held, R.S. Hemler, L.W. Horowitz, S.A. Klein, T.R. Knutson, P.J. Kushner, A.R. Langenhorst, H.-C. Lee, S.-J. Lin, J. Lu, S.L. Malyshev, P.C.D. Milly, V. Ramaswamy, J. Russell, M.D. Schwarzkopf, E. Shevliakova, J.J. Sirutis, M.J. Spelman, W.F. Stern, M. Winton, A.T. Wittenberg, B. Wyman, F. Zeng, and R. Zhang, 2006: GFDL's CM2 global coupled climate models. Part I: Formulation and simulation characteristics. *Journal of Climate*, **19**, 643-674.

Dirmeyer, P.A. and K.L. Brubaker, 2006: Trends in the Northern Hemisphere water cycle. *Geophysical Research Letters*, **33**, L14712, doi:10.1029/2006GL026359.

Doney, S.C., N. Mahowald, I. Lima, R.A. Feely, F.T. Mackenzie, J.-F. Lamarque, and P.J. Rasch, 2007: The impact of anthropogenic atmospheric nitrogen and sulfur deposition on ocean acidification and the inorganic carbon system. *Proceedings of the National Academy of Sciences*, **104**, 14580-14585, doi:10.1073/pnas.0702218104.

Dowsett, H.J., M.A. Chandler, T.M. Cronin, and G.S. Dwyer, 2005: Middle Pliocene sea surface temperature variability. *Paleoceanography*, **20**, PA2014, doi:10.1029/2005PA001133.

Ducklow, H.W., K. Baker, D.G. Martinson, L.B. Quetin, R.M. Ross, R.C. Smith, S.E. Stammerjohn, M. Vernet, and W. Fraser, 2007: Marine pelagic ecosystems: The West Antarctic Peninsula. *Philosophical Transactions of Royal Society B*, **362(1477)**, 67-94, doi:10.1098/rstb.2006.1955.

Erickson, J.E., J.P. Megonigal, G. Peresta, and B.G. Drake, 2007: Salinity and sea level mediate elevated CO_2 effects on C3 and C4 plant interactions and tissue nitrogen in a Chesapeake Bay tidal wetland. *Global Change Biology*, **13**, 202-215, doi:10.1111/j.1365-2486.2006.01285.x.

Euliss, N.H. Jr., R.A. Gleason, A. Olness, R.L. McDougal, H.R. Murkin, R.D. Robarts, R.A. Bourbonniere, and B.G. Warner, 2006: North American prairie wetlands are important nonforested land-based carbon storage sites. *Science of the Total Environment*, **36**, 179-188.

Fan, Y., G. Miguez-Macho, C. Weaver, R. Walko, and A. Robock, 2007: Incorporating water table dynamics in climate modeling, part I: Water table observations and the equilibrium water table. *Journal of Geophysical Research*, **112**, D10125, doi:10.1029/2006JD008111.

Fu, Q. and C.M. Johansson, 2005: Satellite-derived vertical dependence of tropical tropospheric temperature trends. *Geophysical Research Letters*, **32**, L10703, doi:10.1029/2004GL022266.

Fung, I.Y., S.C. Doney, K. Lindsay, and J. John, 2005: Evolution of carbon sinks in a changing climate. *Proceedings of the National Academy of Sciences*, **102(32)**, 11201-11206.

Gettelman, A., E.J. Fetzer, A. Eldering, and F.W. Irion, 2006a: The global distribution of supersaturation in the upper troposphere from the Atmospheric Infrared Sounder. *Journal of Climate*, **19**, 6089-6103.

Gnanadesikan, A., K.W. Dixon, S.M. Griffies, V. Balaji, M. Barreiro, J.A. Beesley, W.F. Cooke, T.L. Delworth, R. Gerdes, M.J. Harrison, I.M. Held, W.J. Hurlin, H.-C. Lee, Z. Liang, G. Nong, R.C. Pacanowski, A. Rosati, J. Russell, B.L. Samuels, Q. Song, M.J. Spelman, R.J. Stouffer, C.O. Sweeney, G. Vecchi, M. Winton, A.T. Wittenberg, F. Zeng, R. Zhang, and J.P. Dunne, 2006: GFDL's CM2 global coupled climate models. Part II: The baseline ocean simulation. *Journal of Climate*, **19**, 675-697.

Gregg, W.W., M.E. Conkright, P. Ginoux, J.E. O'Reilly, and N.W. Casey, 2003: Ocean primary production and climate: Global decadal changes. *Geophysical Research Letters*, **30**, doi:10.1029/2003GL016889.

Guo, L. and R.W. Macdonald, 2006: Source and transport of terrigenous organic matter in the upper Yukon River: Evidence from isotope (δ13C, Δ14C, δ15N) composition of dissolved, colloidal, and particulate phases. *Global Biogeochemical Cycles*, **20**, GB2011, doi:10.1029/2005GB002593.

Han, J. and J.O. Roads, 2004: U.S. climate sensitivity simulated with the NCEP Regional Spectral Model. *Climatic Change*, **62**, 115-154.

Hansen, J. and M. Sato, 2004: Greenhouse gas growth rates. *Proceedings of the National Academy of Sciences*, **101**, 16109-16114, doi:10.1073/pnas.0406982101.

Hayes, D.J. and W.B. Cohen, 2007: Spatial, spectral, and temporal patterns of tropical forest cover change as observed with multiple scales of optical satellite data. *Remote Sensing of Environment*, **106(1)**, 1-16.

Haywood, A.M. and P.J. Valdes, 2004: Modelling Pliocene warmth: contribution of atmosphere, oceans and cryosphere. *Earth and Planetary Science Letters*, **218**, 363-377, doi:10.1016/S0012-821X(03)00685-X.

Haywood, A.M., P. Dekens, A.C. Ravelo, and M. Williams, 2005: Warmer tropics during the mid-Pliocene? Evidence from alkenone paleothermometry and a fully coupled ocean-atmosphere GCM. *Geochemistry, Geophysics, Geosystems*, **6**, Q03010, doi:10.1029/2004GC000799.

Held, I. and B. Soden, 2006: Robust responses of the hydrological cycle to global warming. *Journal of Climate*, **19**, 5686-5699.

IPCC, 2001: *Climate Change 2001: The Scientific Basis — Contribution of Working Group I to the Third Assessment Report* [Houghton, J.T., et al. (eds.)]. Cambridge University Press, Cambridge, UK, 881 pp.

IPCC, 2007a: Climate Change 2007: *The Physical Science Basis. Contribution of Working Group I to the Fourth Assessment Report of the Intergovernmental Panel on Climate Change* [Solomon, S., D. Qin, M. Manning, Z. Chen, M. Marquis, K.B. Averyt, M. Tignor and H.L. Miller (eds.)]. Cambridge University Press, Cambridge, United Kingdom and New York, NY, USA.

IPCC, 2007b: Climate Change 2007: *Impacts, Adaptation and Vulnerability. Contribution of Working Group II to the Fourth Assessment Report of the Intergovernmental Panel on Climate Change* [Parry, M., O. Canziani, J. Palutikof, P. van der Linden, C. Hanson (eds.)]. Cambridge University Press, Cambridge, United Kingdom and New York, NY, USA.

IPCC, 2007c: Climate Change 2007: *Mitigation of Climate Change. Contribution of Working Group III to the Fourth Assessment Report of the Intergovernmental Panel on Climate Change* [Metz, B., O. Davidson, P. Bosch, R. Dave, L. Meyer (eds.)]. Cambridge University Press, Cambridge, United Kingdom and New York, NY, USA.

Jacobs, K., G. Garfin, and M. Lenart, 2005: More than just talk, connecting science and decisionmaking. *Environment*, **47**, 6-21.

Jastrow, J.D., R.M. Miller, R. Matamala, R.J. Norby, T.W. Boutton, C.W. Rice, and C.E. Owensby, 2005: Elevated atmospheric carbon dioxide increases soil carbon. *Global Change Biology*, **11**, 2057-2064.

Jayawickreme, D.H. and D.W. Hyndman, 2007: Evaluating the influence of land cover on seasonal water budgets using NEXRAD rainfall and streamflow data. *Water Resources Research*, **43**, W02408, doi:10.1029/2005WR004460.

Jiang, X., S.J. Eichelberger, D.L. Hartmann, and Y.L. Yung, 2007: Influence of doubled CO_2 on ozone via changes in the Brewer-Dobson Circulation. *Journal of the Atmospheric Sciences*, **64**, 2751-2755.

Kinney, P.L., J.E. Rosenthal, C. Rosenzweig, C. Hogrefe, W. Solecki, K. Knowlton, C. Small, B. Lynn, K. Civerolo, J.-Y. Ku, R. Goldberg, and C. Oliveri, 2006: Assessing potential public health impacts of changing climate and land use: The New York Climate and Health Project. In: *Climate Change and Variability: Impacts and Responses* [Ruth, M., K. Donaghy, and P. Kirshen (eds.)]. New Horizons in Regional Science, Edward Elgar, Cheltenham, UK.

Kueppers, L.M., M.A. Snyder, and L.C. Sloan, 2007: Irrigation cooling effect: Regional climate forcing by land-use change. *Geophysical Research Letters*, **34**, L03703, doi:10.1029/2006GL028679.

Leung, L.R., Y. Qian, X. Bian, W.M. Washington, J. Han, and J.O. Roads, 2004: Mid-century ensemble regional climate change scenarios for the western United States. *Climatic Change*, **62**, 75-113.

Liu, X., S. Xie, and S.J. Ghan, 2007: Evaluation of a new mixed-phase cloud microphysics parameterization with the NCAR single column climate model (SCAM) and ARM M-PACE observations. *Geophysical Research Letters*, **34**, L23712, doi:10.1029/2007GL031446.

Loeb, N.G., B.A. Wielicki, F.G. Rose, and D.R. Doelling, 2006: Variability in global top-of-atmosphere shortwave radiation between 2000 and 2005. *Geophysical Research Letters*, **34**, L03704, doi:10.1029/2006GL028196.

Logan, J.A. and J.A. Powell, 2007: Ecological consequences of forest-insect disturbance altered by climate change. In: *Climate Change in Western North America: Evidence and Environmental Effects* [Wagner, F.H. (ed.)]. University of Utah Press, Salt Lake City, Utah (in press).

Lubin, D. and A.M. Vogelmann, 2006: A climatologically significant aerosol longwave indirect effect in the Arctic. *Nature*, **439**, 453-456, doi:10.1038/nature04449.

Luo, Y., D. Hui, and D. Zhang, 2006: Elevated carbon dioxide stimulates net accumulations of carbon and nitrogen in terrestrial ecosystems: A meta-analysis. *Ecology*, **87**, 53-63.

MEA, 2005: *Ecosystems and Human Well-being: Synthesis.* Millennium Ecosystem Assessment, Island Press, Washington, DC, USA, 137 pp.

Mace, G.G., S. Benson, K. Sonntag, S. Kato, Q. Min, P. Minnis, C. Twohy, M. Poellot, X. Dong, C. Long, Q. Zhang, and D. Doelling, 2006a: Cloud radiative forcing at the ARM Climate Research Facility: Part 1. Technique, validation, and comparison to satellite-derived diagnostic quantities. *Journal of Geophysical Research*, **111**, D11S90, doi:10.1029/2005JD005921.

Mace, G.G., S. Benson, and S. Kato, 2006b: Cloud radiative forcing at the ARM Climate Research Facility: Part 2. The vertical redistribution of radiant energy by clouds. *Journal of Geophysical Research*, **111**, D11S91, doi:10.1029/2005JD005922.

Marshall, P.A. and H.Z. Schuttenberg, 2006: *A Reef Manager's Guide to Coral Bleaching.* Great Barrier Reef Marine Park Authority, Townsville, Australia, 163 pp.

Mason, S.J., 2004: Simulating climate over western North America using stochastic weather generators. *Climatic Change*, **62**, 155-187.

Meehl, G.A. and C. Tebaldi, 2004: More intense, more frequent and longer lasting heat waves in the 21st century. *Science*, **305**, 994-997.

Meehl, G.A., W.M. Washington, C. Ammann, J.M. Arblaster, T.M.L. Wigley, and C. Tebaldi, 2004a: Combinations of natural and anthropogenic forcings and 20th century climate. *Journal of Climate*, **17**, 3721-3727.

Meehl, G.A., C. Covey, and M. Latif, 2004b: Soliciting participation in climate model analyses leading to IPCC Fourth Assessment Report. *Eos, Transactions of the American Geophysical Union*, **85(29)**, 274, doi:10.1029/2004EO290002.

Meehl, G.A., C. Covey, B. McAvaney, M. Latif, and R.J. Stouffer, 2005: Overview of the Coupled Model Intercomparison Project. *Bulletin of the American Meteorological Society*, **86**, 89-93.

Melack, J.M., L.L. Hess, M. Gastil, B.R. Forsberg, S.K. Hamilton, I.B.T. Lima, and E.M.L.M. Novo, 2004: Regionalization of methane emissions in the Amazon Basin with microwave remote sensing. *Global Change Biology*, **10**, 530-544.

Monaghan, A.J., D.H. Bromwich, R.L. Fogt, S.-H. Wang, P.A. Mayewski, D.A. Dixon, A. Ekaykin, M. Frezzotti, I. Goodwin, E. Isaksson, S.D. Kaspari, V.I. Morgan, H. Oerter, T.D. Van Ommen, C.J. Van der Veen, and J. Wen, 2006: Insignificant change in Antarctic snowfall since the International Geophysical Year. *Science*, **313**, 827-831.

Mote, P.W., A.F. Hamlet, M.P. Clark, and D.P. Lettenmaier, 2005: Declining mountain snowpack in western North America. *Bulletin of the American Meteorological Society*, **86**, 39-49.

Muller-Karger, F.E., R. Varela, R. Thunell, R. Luerssen, C. Hu, and J.J. Walsh, 2005: The importance of continental margins in the global carbon cycle. *Geophysical Research Letters*, **32**, L01602, doi:10.1029/2004GL021346.

NRC, 1999: *Capacity of U.S. Climate Modeling to Support Climate Change Assessment Activities*. Climate Research Committee, National Research Council, National Academy Press, Washington, DC.

NRC, 2000: *Reconciling Observations of Global Temperature Change*. National Research Council, National Academy Press, Washington, DC, USA, 78 pp.

NRC, 2001a: *Climate Change Science: An Analysis of Some Key Questions*. Committee on the Science of Climate Change, National Research Council, National Academy Press, Washington, DC, USA, 42 pp.

NRC, 2001b: *Improving the Effectiveness of U.S. Climate Modeling*. Panel on Improving the Effectiveness of U.S. Climate Modeling, Board on Atmospheric Sciences and Climate, National Research Council, National Academy Press, Washington, DC.

NRC, 2002: *Abrupt Climate Change: Inevitable Surprises*. National Research Council, National Academy Press, Washington, DC, USA, 230 pp.

NRC, 2005: *Thinking Strategically: The Appropriate Use of Metrics for the Climate Change Science Program*. National Research Council, National Academy Press, Washington, DC.

NRC, 2007a: *Evaluating Progress of the U.S. Climate Change Science Program: Methods and Preliminary Results*. National Research Council, National Academy Press, Washington, DC.

NRC, 2007b: *Analysis of Global Change Assessments: Lessons Learned*. National Research Council, National Academy Press, Washington, DC.

NRC, 2007c: *Earth Science and Applications from Space: National Imperatives for the Next Decade and Beyond*. National Research Council, National Academy Press, Washington, DC, USA, 400 pp.

NSTC, 2007: *Charting the Course for Ocean Science in the United States for the Next Decade: An Ocean Research Priorities Plan and Implementation Strategy*. National Science and Technology Council Joint Subcommittee on Ocean Science and Technology, Washington, DC, 84 pp. Available at <ocean.ceq.gov/about/docs/orppfinal.pdf>.

Newman, P.A., E.R. Nash, S.R. Kawa, S.A. Montzka, and S.M. Schauffler, 2006: When will the Antarctic ozone hole recover? *Geophysical Research Letters*, **33**, doi:10.1029/2005GL025232.

Norby, R.J., E.H. DeLucia, B. Gielen, C. Calfapietra, C.P. Giardina, J.S. King, J. Ledford, H.R. McCarthy, D.J.P. Moore, R. Ceulemans, P. De Angelis, A.C. Finzi, D.F. Karnosky, M.E. Kubiske, M. Lukac, K.S. Pregitzer, G.E. Scarascia-Mugnozza, W.H. Schlesinger, and R. Oren,

2005: Forest response to elevated CO_2 is conserved across a broad range of productivity. *Proceedings of the National Academy of Sciences*, **102**, 18052-18056.

Orr, J.C., V.J. Fabry, O. Aumont, L. Bopp, S.C. Doney, R.A. Feely, A. Gnanadesikan, N. Gruber, A. Ishida, F. Joos, R.M. Key, K. Lindsay, E. Maier-Reimer, R. Matear, P. Monfray, A. Mouchet, R.G. Najjar, G.-K. Plattner, K.B. Rodgers, C.L. Sabine, J.L. Sarmiento, R. Schlitzer, R.D. Slater, I.J. Totterdell, M.-F. Weirig, Y. Yamanaka, and A. Yool, 2005: Anthropogenic ocean acidification over the twenty-first century and its impact on calcifying organisms. *Nature*, **437**, 681-686.

Ovtchinnikov, M., T. Ackerman, R. Marchand, and M. Khairoutdinov, 2006: Evaluation of the multiscale modeling framework using data from the Atmospheric Radiation Measurement Program. *Journal of Climate*, **19**, 1716-1729, doi:10.1175/JCLI3699.1.

Pielke, R.A., C. Landsea, M. Mayfield, J. Laver, and R. Pasch, 2005: Hurricanes and global warming. *Bulletin of the American Meteorological Society*, **86**, 1571-1575.

Potter, C., S. Klooster, S. Hiatt, M. Fladeland, V. Genovese, and P. Gross, 2007: Satellite-derived estimates of potential carbon sequestration through afforestation of agricultural lands in the United States. *Climatic Change*, **80**, 323-336, doi:10.1007/s10584-006-9109-3.

Poumadere, M., C. Mays, S. LeMer, and R. Blong, 2005: The 2003 heat wave in France: Dangerous climate change here and now. *Risk Analysis*, **25**, 1483-1494.

Randerson, J.T., H. Liu, M.G. Flanner, S.D. Chambers, Y. Jin, P.G. Hess, G. Pfister, M.C. Mack, K.K. Treseder, L.R. Welp, F.S. Chapin, J.W. Harden, M.L. Goulden, E. Lyons, J.C. Neff, E.A.G. Schuur, and C.S. Zender, 2006: The impact of boreal forest fire on climate warming. *Science*, **314**, 1130-1132, doi:10.1126/science.1132075.

Rignot, E. and P. Kanagaratnam, 2006: Changes in the velocity structure of the Greenland Ice Sheet. *Science*, **311**, 986.

Rosen, R., A. Chu, J.J. Szykman, R. DeYoung, J.A. Al-Saadi, A. Kaduwela, and C. Bohnenkamp, 2006: Application of satellite data for three-dimensional monitoring of PM2.5 formation and transport in San Joaquin Valley, California. In: *Remote Sensing of Aerosol and Chemical Gases, Model Simulation/Assimilation, and Applications to Air Quality* [Chu, A., J. Szykman, and S. Kondragunta (eds.)]. Proceedings of SPIE - International Society of Optical Engineering, **6299**, doi:10.1117/12.681649.

Sarmiento, J.L., S.C. Wofsy, and the members of the Carbon and Climate Working Group, 1999: *A U.S. Carbon Cycle Science Plan*. U.S. Global Change Research Program, Washington, DC, USA, 69 pp.

Schmidt, G.A. and 35 other authors, 2006: Present-day atmospheric simulations using GISS ModelE: Comparison to in situ, satellite, and reanalysis data. *Journal of Climate*, **19**, 153-192.

Schmittner, A., 2005: Decline of the marine ecosystem caused by a reduction in the Atlantic overturning circulation. *Nature*, **434**, 628-633.

Seager, R., Y. Kushnir, C. Herweijer, N. Naik, and J. Velez, 2005: Modeling of tropical forcing of persistent droughts and pluvials over western North America: 1856-2000. *Journal of Climate*, **18**, 4065-4088.

Sekercioglu, C.H., G.C. Daily, and P.R. Ehrlich, 2004: Ecosystem consequences of bird declines. *Proceedings of the National Academy of Sciences*, **101**, 18042-18047.

Soden, B.J. and I.M. Held, 2006: An assessment of climate feedbacks in coupled ocean-atmosphere models. *Journal of Climate*, **19**, 3354-3360.

Stirling, I. and C.L. Parkinson, 2006: Possible effects of climate warming on selected populations of polar bears in the Canadian Arctic. *Arctic*, **59(3)**, 261-275.

Stolarski, R.S., A.R. Douglass, M. Gupta, P.A. Newman, S. Pawson, M.R. Schoeberl, and J.E. Nielsen, 2006: An ozone increase in the Antarctic summer stratosphere: A dynamical response to the ozone hole. *Geophysical Research Letters*, **33**, L21805, doi:10.1029/2006GL026820.

Stouffer, R.J., J. Yin, J.M. Gregory, K.W. Dixon, M.J. Spelman, W. Hurlin, A.J. Weaver, M. Eby, G.M. Flato, H. Hasumi, A. Hu, J.H. Jungclaus, I.V. Kamenkovich, A. Levermann, M. Montoya, S. Murakami, S. Nawrath, A. Oka, W.R. Peltier, D.Y. Robitaille, A. Sokolov, G. Vettoretti, and S.L. Weber, 2006a: Investigating the causes of the response of the thermohaline circulation to past and future climate changes. *Journal of Climate*, **19**, 1365-1387.

Stouffer, R.J., T.L. Delworth, K.W. Dixon, R. Gudgel, I. Held, R. Hemler, T. Knutson, M.D. Schwarzkopf, M.J. Spelman, M.W. Winton, A.J. Broccoli, H-C. Lee, F. Zeng, and B. Soden, 2006b: GFDL's CM2 global coupled climate models. Part IV: Idealized climate response. *Journal of Climate*, **19**, 723-740.

Sun, G., C. Li, C.C. Trettin, J. Lu, and S.G. McNulty, 2006: Simulating the biogeochemical cycles in cypress wetland pine upland ecosystems at a landscape scale with the wetland-DNDC model. In: *Hydrology and Management of Forested Wetlands, Proceedings of the International Conference 8-12 April 2006*. American Society of Agricultural and Biological Engineers, St. Joseph, MI.

Thomson, M.C., F.J. Doblas-Reyes, S.J. Mason, R. Hagedorn, R.J. Connor, T. Phindela, A.P. Morse, and T.N. Palmer, 2006: Malaria early warnings based on seasonal climate forecasts from multimodel ensembles. *Nature*, **439**, 576-579.

Turner, D.D., 2007: Improved ground-based liquid water path retrievals using a combined infrared and microwave approach. *Journal of Geophysical Research*, **112**, D15204, doi:10.1029/2007JD008530.

Turner, D.D., A.M. Vogelmann, R. Austin, J.C. Barnard, K. Cady-Pereira, C. Chiu, S.A. Clough, C.J. Flynn, M.M. Khaiyer, J.C. Liljegren, K. Johnson, B. Lin, C.N. Long, A. Marshak, S.Y. Matrosov, S.A. McFarlane, M.A. Miller, Q. Min, P. Minnis, W. O'Hirok, Z. Wang, and W. Wiscombe, 2007: Thin liquid water clouds: Their importance and our challenge. *Bulletin of the American Meteorological Society*, **88**, 177-190.

Urbanski, S., C. Barford, S. Wofsy, C. Kucharik, E. Pyle, J. Budney, K. McKain, D. Fitzjarrald, M. Czikowsky, and J.W. Munger, 2007: Factors controlling CO_2 exchange on time scales from hourly to decadal at Harvard Forest. *Journal of Geophysical Research*, **112**, G02020, doi:10.1029/2006JG000293.

USCLIVAR, 2002: *Climate Process Modeling and Science Teams (CPTs): Motivation and Concept*. Report 2002-1, Scientific Steering Committee, U.S. CLIVAR Office, Washington, DC, USA, 4 pp.

Velicogna, I. and J. Wahr, 2005: Greenland mass balance from GRACE. *Geophysical Research Letters*, **32**, L18505, doi:10.1029/2005GRL023955.

Velicogna, I. and J. Wahr, 2006: Measurements of time-variable gravity show mass loss in Antarctica. *Science*, **311**, 1754-1756.

WMO, 2003: *Scientific Assessment of Ozone Depletion 2002*. Global Ozone Research and Monitoring Project Report No. 47, World Meteorological Organization, Geneva, 498 pp.

Wittenberg, A.T., A. Rosati, N.-C. Lau, and J.J. Ploshay, 2006: GFDL's CM2 global coupled climate models. Part III: Tropical Pacific climate and ENSO. *Journal of Climate*, **19**, 698-722.

Wood, A.W., L.R. Leung, V. Sridhar, and D.P. Lettenmaier, 2004: Hydrologic implications of dynamical and statistical approaches to downscaling climate model outputs. *Climatic Change*, **62**, 189-216

Woods, T.N. and J. Lean, 2007: Anticipating the next decade of Sun-Earth system variations. *Eos*, **88(44)**, 457-458.

Yu, H., Y.J. Kaufman, M. Chin, G. Feingold, L.A. Remer, T.L. Anderson, Y. Balkanski, N. Bellouin, O. Boucher, S. Christopher, P. DeCola, R. Kahn, D. Koch, N. Loeb, M.S. Reddy, M. Schultz, T. Takemura, and M. Zhou, 2006: A review of measurement-based assessments of the aerosol direct radiative effect and forcing. *Atmospheric Chemistry and Physics*, **6**, 613-666.

Zhou, G., S. Liu, Z. Li, D. Zhang, X. Tang, C. Zhou, J. Yan, and J. Mo, 2006: Old-growth forests can accumulate carbon in soils. *Science*, **314**, 1417, doi:10.1126/science.1130168.

Zimov, S.A., E.A.G. Schuur, and F.S. Chapin III, 2006: Permafrost and the global carbon budget. *Science*, **312**, 1612-1613.

Ziska, L.H. and G.B. Runion, 2006: Rising atmospheric carbon dioxide and global climate change: Assessing the potential impact on agro-ecosystems by weeds, insects, and diseases. In: *Agroecosystems in a Changing Climate* [Newton, P.C.D., A. Carran, G.R. Edwards, and P.A. Niklaus (eds.)]. CRC Press, Boston, MA, USA, Chapter 11, pp. 261-287.

Appendices

Appendix 1: CCSP Synthesis and Assessment Products

See <www.climatescience.gov/Library/sap/sap-summary.php> for current information regarding lead and contributing agencies, production, and publication scheduling.

CCSP Goal 1: Improve knowledge of the Earth's past and present climate and environment, including its natural variability, and improve understanding of the causes of observed variability and change.

SAP 1.1 – TEMPERATURE TRENDS IN THE LOWER ATMOSPHERE: STEPS FOR UNDERSTANDING AND RECONCILING DIFFERENCES

Temperature change is a fundamental measure of climate change. This product, which was the first to be completed, addresses temperature changes from the surface through the lower stratosphere and understanding of the causes of these changes. It assesses progress made since the reports by the National Research Council (NRC, 2000) and the Intergovernmental Panel on Climate Change (IPCC, 2001) and highlights differences between the individual temperature records obtained by components of the existing observational and modeling systems and documents the potential causes of these differences.

SAP 1.2 – PAST CLIMATE VARIABILITY AND CHANGE IN THE ARCTIC AND AT HIGH LATITUDES

The Arctic and the high latitudes have warmed more rapidly than almost any other region on Earth over at least the last millennium. This warming has been accompanied by a decrease in sea ice cover and thickness and a decrease in ocean salinity. In addition, significant changes in the permafrost active layer are now being detected. The impacts on humans and ecosystems that are associated with these changes were reported in the Arctic Climate Impact Assessment, which was partially funded by CCSP-participating agencies. The present synthesis and assessment product on the Arctic and high latitudes will focus on the state of knowledge concerning past changes in the physical climate of this region and the implications of this record of past changes for current and future change. This information is vital since high-latitude regions are projected to continue to experience the greatest warming in the future.

SAP 1.3 – RE-ANALYSES OF HISTORICAL CLIMATE DATA FOR KEY ATMOSPHERIC FEATURES: IMPLICATIONS FOR ATTRIBUTION OF CAUSES OF OBSERVED CHANGE

A reanalysis is a detailed, retrospective study of the state of the atmosphere using a consistent numerical model of the dynamics of the system and based on observations for the time period of the study. This product will provide an assessment of the capability and limitations of state-of-the-art climate reanalysis to describe past and current climate conditions, and the consequent implications for scientifically interpreting the causes of climate variations and change. The product will be in the form of a report that summarizes the present status of national and international climate reanalysis efforts, and discusses key research findings on the strengths and limitations of current reanalysis products for describing and analyzing the causes of climate variations and trends that have occurred during the time period of the reanalysis records (roughly the past half-century). The report will describe how reanalysis products have been used in documenting, integrating, and advancing knowledge of climate system behavior, as well as in ascertaining significant remaining uncertainties in descriptions and physical understanding of the climate system.

CCSP Goal 2: Improve quantification of the forces bringing about changes in the Earth's climate and related systems.

SAP 2.1 – SCENARIOS OF GREENHOUSE GAS EMISSIONS AND ATMOSPHERIC CONCENTRATIONS AND REVIEW OF INTEGRATED SCENARIO DEVELOPMENT AND APPLICATION

This product, released in 2007, provides a new long-term, global reference for greenhouse gas stabilization scenarios and an evaluation of the process by which scenarios are developed and used. SAP 2.1 consists of two parts. Part A, *Scenarios of Greenhouse Gas Emissions and Atmospheric Concentrations*, uses computer-based scenarios to evaluate four alternative stabilization levels of greenhouse gases in the atmosphere and the implications for energy and the economy of achieving each level. Part A includes stabilization scenarios for the six primary anthropogenic greenhouse gases – carbon dioxide, nitrous oxide, methane, hydrofluorocarbons, perfluorocarbons, and sulfur hexafluoride – and it uses updated economic and technological data and new tools for scenario development. Although these scenarios should not be considered definitive predictions of future events, they provide valuable insights for decisionmakers. Part B, *Global Change Scenarios: Their Development and Use*, examines how scenarios have been developed and used in global climate change applications, evaluates the effectiveness of current scenarios, and recommends ways to make future scenarios more useful. Part B of the report concludes that scenarios can support decisionmaking by providing insights regarding key uncertainties, including future emissions and climate as well as other environmental and economic conditions.

SAP 2.2 – NORTH AMERICAN CARBON BUDGET AND IMPLICATIONS FOR THE GLOBAL CARBON CYCLE

This product provides a synthesis and integration of the current knowledge of the North American carbon budget (including land, atmosphere, inland waters, and adjacent oceans) and its context within the global carbon cycle. In a format useful to decisionmakers, it summarizes knowledge of carbon cycle properties and changes relevant to the contributions of, and impacts on, the United States and the rest of the world and provides scientific information for U.S. decision support focused on key issues for carbon management and policy. It addresses carbon emissions; natural reservoirs and sequestration; rates of transfer; the consequences of changes in carbon cycling; effects of purposeful carbon management; effects of agriculture, forestry, and natural resource management; and socioeconomic drivers and consequences. The report includes an analysis of North America's carbon budget that documents the state of knowledge and quantifies uncertainties.

SAP 2.3 – AEROSOL PROPERTIES AND THEIR IMPACTS ON CLIMATE

Aerosols can cause a net cooling or warming within the climate system, depending upon their physical and chemical characteristics. In addition to these direct effects, aerosols can also have indirect effects on radiative forcing of the climate system by changing cloud properties. The first phase of development of this product is to produce major scientific reviews of the following three topics: dependence of radiative forcing by tropospheric

aerosols on aerosol composition in the north Atlantic, Pacific, and Indian Ocean regions; measurement-based understanding of aerosol radiative forcing from remote-sensing observations; and model intercomparison to quantify uncertainties associated with indirect aerosol forcing. The second-phase product will draw upon the scientific information gathered by the development of the IPCC Fourth Assessment Report and the NRC review, *Radiative Forcing of Climate Change*. These community-wide assessments of climate change (and the aerosol-climate topic inclusively) will be drawn from in writing SAP 2.3.

SAP 2.4 – TRENDS IN EMISSIONS OF OZONE-DEPLETING SUBSTANCES, OZONE LAYER RECOVERY, AND IMPLICATIONS FOR ULTRAVIOLET RADIATION EXPOSURE

Measurements of ozone-depleting gases in the atmosphere have shown that the concentrations of these gases are declining in response to the agreements reached under the Montreal Protocol. This report will provide an update on trends in stratospheric ozone, ozone-depleting gases, and ultraviolet radiation exposure; progress in improving model evaluations of the sensitivity of the ozone layer to changes in atmospheric composition and climate; and relevant implications for the United States. This information is key in ensuring that international agreements to phase out production of ozone-depleting substances are having the expected outcome – recovery of the protective ozone layer. The report will derive most of its information from recent international assessments of stratospheric ozone, ozone-depleting substances, and climate.

CCSP Goal 3: Reduce uncertainty in projections of how the Earth's climate and environmental systems may change in the future.

SAP 3.1 – CLIMATE CHANGE MODELS: AN ASSESSMENT OF STRENGTHS AND LIMITATIONS

The topics addressed by this product are the strengths and limitations of climate models at different spatial and temporal scales. Its purpose is to provide information on the results from climate models, in ways that will allow the potential user of the information to evaluate how best it may be applied. The product will focus on natural and human-caused factors influencing climate variability and change during the period from 1870 to 2000. It will characterize sources of uncertainty in climate models and their implications for estimating future climate change. This product will be limited to the models and their sensitivity, feedbacks, strengths, and limitations, rather than making specific future projections.

SAP 3.2 – CLIMATE PROJECTIONS FOR RESEARCH AND ASSESSMENT BASED ON EMISSIONS SCENARIOS DEVELOPED THROUGH THE CCTP

This product will have two distinct components. The first will be to produce climate projections for research and assessment based on greenhouse gas emission scenarios and atmospheric concentrations as reported in SAP 2.1a. The second will be to assess the future climate impacts of short-lived gaseous and particulate species.

SAP 3.3 – WEATHER AND CLIMATE EXTREMES IN A CHANGING CLIMATE: REGIONS OF FOCUS - NORTH AMERICA, HAWAII, CARIBBEAN, AND U.S. PACIFIC ISLANDS

The impact of climate extremes can be severe and wide-ranging. There is evidence that the economic impact of weather and climate extremes in the United States has increased over the past several decades, but the evidence for increases in extreme weather and climate events varies depending on the event of interest. These events may be related to temperature parameters (severe freezes, heat waves), precipitation (wet spells, heavy precipitation events, droughts, ice and hail, snow cover and depth), or tropical and extratropical storm frequency. Identifying recent changes and trends in such parameters will be a focus of the report, as well as identifying what can be said about future changes. Since extreme weather and climate events on a global scale are regularly addressed in international assessments, this product will focus on weather and climate extremes primarily across Canada, Mexico, and the United States.

SAP 3.4 – ABRUPT CLIMATE CHANGE

The paleoclimate record reveals that Earth's climate can change rapidly and strongly between different stable states. Various scenarios portray future abrupt climate change large enough to pose a significant challenge to society. The goal of this product is to review and synthesize current understanding of abrupt climate change and to identify gaps in knowledge. The report will integrate information from the paleoclimate record, the instrumental record, and numerical model-based studies at various spatial scales. Key identified risks, such as changes in ocean thermohaline circulation and alteration of terrestrial hydrologic conditions (e.g., the location or amount of

precipitation) will receive special attention because the potential impacts on society are large.

CCSP Goal 4: Understand the sensitivity and adaptability of different natural and managed ecosystems and human systems to climate and related global changes.

SAP 4.1 – COASTAL ELEVATION AND SENSITIVITY TO SEA-LEVEL RISE

This product will examine the vulnerability of coastal areas in the U.S. mid-Atlantic states to sea-level change. Specific questions to be addressed include identifying which areas are low enough to be inundated by tides, how floodplains would change due to a changing climate, which areas might be subject to erosion, and locations where wetlands will be able to migrate inland versus locations where shores will be protected. The product will examine the implications of sea-level rise, including impacts on population and economic activity in vulnerable areas, costs of shore protection, ecological effects, flood damages, public access to modified shore areas, cases where sea-level rise justifies policy changes, options being considered by conservancies and governments, and lessons from the unfolding consequences of the 2005 hurricanes in the Gulf Coast region.

SAP 4.2 – THRESHOLDS OF CHANGE IN ECOSYSTEMS

There is a body of ecosystems research that focuses on enhancing understanding of climate change impacts on ecosystems (and *vice versa*) and developing the capability to predict potential impacts of future climate change. Increasing emphasis is being placed on climate-related thresholds that could result in discontinuities or sudden changes in ecosystems and climate-sensitive resources. Discontinuities in responses of ecosystems and resources are difficult to predict, and may significantly affect human societies that depend on ecosystem goods and services. Improved understanding of such sudden changes is essential to managing ecosystems and resources in the face of climate change. This report will synthesize the present state of scientific understanding regarding thresholds of change that trigger sudden changes in ecosystems and climate-sensitive resources. The report will develop a conceptual framework for characterizing sudden changes, and synthesize peer-reviewed studies that provide the best available evidence for defining circumstances that trigger discontinuities in response to climate change.

SAP 4.3 – THE EFFECTS OF CLIMATE CHANGE ON AGRICULTURE, BIODIVERSITY, LAND, AND WATER RESOURCES

This report addresses the effects of climate change on agriculture, forestry, land and water resources, and biodiversity. Air and water temperature, precipitation, and related climate variables are fundamental regulators of biological processes. For this reason, human-induced climate change has the potential to affect the condition, composition, structure, and function of ecosystems. Such changes may also alter the linkages and feedbacks between ecosystems and the climate system. Additionally, ecosystems produce a wide array of goods and services valued by humans and in many cases essential for human survival and property. Climate-related changes in ecosystems and other key resources could have impacts on human communities and economic conditions.

SAP 4.4 – PRELIMINARY REVIEW OF ADAPTATION OPTIONS FOR CLIMATE-SENSITIVE ECOSYSTEMS AND RESOURCES

Climate is a dominant factor influencing the distribution, abundance, structure, and function of, and services provided by, ecosystems. Many ecosystems are thus vulnerable to future changes in climate. The goal of adaptation is to reduce these risks of adverse ecological outcomes through management activities that increase the resilience of these systems to climate change. Resilience is defined here as the magnitude of disturbance that can be absorbed by a system before it shifts from one stable state (or stability domain) to another and the speed of return of a system to equilibrium after a disturbance has occurred. This report provides a review and synthesis of information on adaptation options for selected climate-sensitive ecosystems in order to aid in designing management strategies that facilitate adaptation, provides examples of how to implement strategies in specific places, and identifies issues and challenges associated with implementation of adaptation options.

SAP 4.5 – EFFECTS OF CLIMATE CHANGE ON ENERGY PRODUCTION AND USE IN THE UNITED STATES

This report summarizes what is currently known about potential effects of climatic change on energy production and use in the United States. It focuses on three questions: (1) How might climatic change affect energy use in the

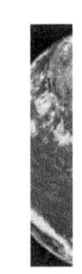

United States, (2) how might climatic change affect energy production and supply in the United States, and (3) how might climatic change have other effects that indirectly shape energy production and use in the United States? Great care was taken in answering these questions, for two reasons. One, the available research literature on these key questions is limited, supporting a discussion of issues but not providing definite answers. Two, as with many other aspects of potential effects of climatic change on the United States, the effects on energy production and use depend on more than climatic change alone; other potentially important factors include patterns of economic growth and land use, patterns of population growth and distribution, technological change, and social and cultural trends that could shape policies and actions, individually and institutionally.

SAP 4.6 – ANALYSES OF THE EFFECTS OF GLOBAL CHANGE ON HUMAN HEALTH AND WELFARE AND HUMAN SYSTEMS

This product examines the effects of global change on human systems. The impacts of climate variability, climate change, shifting patterns of land use, and changes in population patterns are human problems, not simply problems for the natural or the physical world. This SAP examines the vulnerability of human health and socioeconomic systems to global environmental change across three areas of potential impacts and adaptations: human health, human settlements, and human welfare. It addresses the questions of how, where, and when climate variability and change will affect U.S. social systems. The challenge for this project was to assess risks associated with health, welfare, and settlements and to identify and develop timely adaptive strategies to address human vulnerabilities. The primary goals for adaptation to climate change and variability focus on managing significant risks proactively when possible; establishing protocols to detect and measure risks; and leveraging technical and institutional adaptive capacity to address new climate risks, especially as they exceed conventional adaptive measures.

SAP 4.7 – IMPACTS OF CLIMATE CHANGE AND VARIABILITY ON TRANSPORTATION SYSTEMS AND INFRASTRUCTURE: GULF COAST STUDY

This product addresses the potential effects of climate variability and change on transportation infrastructure and systems in the central Gulf Coast of the United States. The purpose of this study was to increase the knowledge base regarding the risks and sensitivities of transportation infrastructure to climate variability and change, the significance of these risks, and the range of adaptation strategies that may be considered to ensure a robust and reliable transportation network. Implications for all transportation modes – surface, marine, and aviation – are addressed. The three-phase study focuses on the Gulf Coast, and assesses the significant risks to transportation, develops methodology to be applied in other geographic locations, identifies potential strategies for adaptation, and develops decision-support tools to assist transportation decisionmakers in incorporating climate-related trend information into transportation system planning, design, engineering, and operational decisions.

CCSP Goal 5: Explore the uses and identify the limits of evolving knowledge to manage risks and opportunities related to climate variability and change.

SAP 5.1 – USES AND LIMITATIONS OF OBSERVATIONS, DATA, FORECASTS, AND OTHER PROJECTIONS IN DECISION SUPPORT FOR SELECTED SECTORS AND REGIONS

The product will focus on characterizing a subset of the observations from remote-sensing and *in situ* instrumentation that is of high value for decisionmaking. The product will characterize observational capabilities that are currently or could potentially be used in decision-support tools, catalog a subset of ongoing decision-support activities that use these capabilities, and evaluate a limited number of case studies of these decision-support activities. The detailed evaluation of decision-support activities and demonstration projects will provide information to agencies and organizations responsible for developing, operating, and maintaining selected decision-support processes and tools. The evaluation will also provide information on the nature of interactions between users and producers of climate science information, approaches for accessing science information, and assimilation of scientific information in the decisionmaking process. The product will include an on-line catalog of decision-support demonstration projects with interactive links, which will

be updated as additional experiments are conducted and new approaches to incorporating and benchmarking the application of observations and other global change research products evolve.

SAP 5.2 – BEST PRACTICE APPROACHES FOR CHARACTERIZING, COMMUNICATING, AND INCORPORATING SCIENTIFIC UNCERTAINTY IN DECISIONMAKING

This product will address the issue of uncertainty and its relationship to science, assessment, and decisionmaking. Specifically, the product is intended to help improve the quality and consistency of information about scientific uncertainty presented to decisionmakers and other users of CCSP reports by identifying 'best practice' options recommended in the literature on this subject; to improve communication between scientists and users of the products by providing recommendations for addressing uncertainty; and to provide a brief overview of the literature on approaches for communicating and considering uncertainty related to climate.

SAP 5.3 – DECISION-SUPPORT EXPERIMENTS AND EVALUATIONS USING SEASONAL TO INTERANNUAL FORECASTS AND OBSERVATIONAL DATA

This product will concentrate on the water resource management sector. It will describe and evaluate current forecasts, assess how forecasts are being used in decision settings, and evaluate decisionmakers' level of confidence in these forecasts. The participants in the development of this product (primarily consisting of government officials, researchers, and users) will evaluate forecasts as well as their delivery, to identify options for improving partnerships between the research and user communities. It will inform decisionmakers about the experiences of others who have experimented with the use of seasonal and interannual forecasts and other observational data; inform climatologists and social scientists about how to advance the delivery of decision-support resources that use the most recent forecast products, methodologies, and tools; and inform science managers as they plan for future investments in research related to forecasts and their role in decision support.

Appendix 2: Selected Observing Missions and Networks and Data Management Systems Relevant to CCSP Research

The United States is contributing to the development and operation of several global observing systems that collectively attempt to combine data streams from both research and operational observing platforms to provide a comprehensive measure of climate system variability and climate change processes. These systems provide a baseline Earth-observing system and include National Aeronautics and Space Administration (NASA), National Oceanic and Atmospheric Administration (NOAA), and U.S. Geological Survey (USGS) Earth-observing satellites and extensive *in situ* observational capabilities. CCSP also supports several ground-based measurement activities that provide the data used in studies of the various climate processes necessary for better understanding of climate change.

U.S. observational, monitoring, and data management activities contribute significantly to several international observing systems including the Global Climate Observing System principally sponsored by the World Meteorological Organization (WMO); the Global Ocean Observing System sponsored by the United Nations Educational, Scientific, and Cultural Organization's Intergovernmental Oceanographic Commission (IOC); and the Global Terrestrial Observing System sponsored by the United Nations Food and Agriculture Organization. The latter two have climate-related elements being developed jointly with the Global Climate Observing System. The United States is also playing an important role in the Global Earth Observation System of Systems (GEOSS), which is an international framework for coordinating and sustaining the aforementioned (and other) systems. The following list is not intended to be exhaustive, but to provide examples of key observing systems and networks and data management systems relevant to CCSP research.

Selected Missions, Observing Capabilities, and Networks

AmeriFlux. Based on systematic eddy-covariance measurements, AmeriFlux Network sites provide data on exchange of carbon dioxide (CO_2), water vapor, and energy between air and plant canopies for a range of different types of ecosystems. Time averages of these continuous measurements are expressed as net ecosystem exchange (NEE) for any selected time period, and the annual CO_2 NEE, for example, is often termed net ecosystem production (NEP), or the quantity of carbon gained or lost by the system. Some AmeriFlux data products also include measures of respiration, which when combined with NEE (CO_2) enable estimates of gross primary production (GEP). Both NEP and GEP are important for calculating terrestrial carbon budgets and for carbon cycle analysis and modeling. AmeriFlux data products also include a number of corollary biometric and micrometeorological measurements that are used to understand carbon cycle processes. See <public.ornl.gov/ameriflux/>.

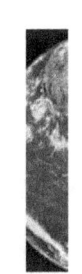

Aquarius. Aquarius is a satellite mission to measure global sea surface salinity. The average ocean salinity is about 35 parts per 1,000. The instruments that are part of this satellite mission will measure changes in sea surface salinity over the global oceans to a precision of 0.2 parts per 1,000 (equivalent to about one-sixth of a teaspoon of salt in one gallon of water). By measuring global sea surface salinity with high spatial and temporal resolution, Aquarius will answer long-standing questions about how oceans respond to climate change and the water cycle, including changes in freshwater input and output to the ocean associated with precipitation, evaporation, ice melting, and river runoff. Aquarius is a collaboration between NASA and CONAE, the Argentine space agency, with an expected launch date in 2010.

The Arctic Observing Network. The Arctic Observing Network (AON), developed largely as a research system by the National Science Foundation (NSF) and NOAA, is envisioned as a system of atmospheric, land, and ocean-based environmental monitoring capabilities — from ocean buoys to satellites — that will significantly advance observations of Arctic environmental conditions in order to better understand the wide-ranging series of significant and rapid changes occurring in the Arctic. In 2008, aircraft flights over the North Slope of Alaska will measure temperature, humidity, total particle number, aerosol size distribution, cloud condensation nuclei concentration, ice nuclei concentration, optical scattering and absorption, vertical velocity, cloud liquid water and ice contents, cloud droplet and crystal size distributions, cloud particle shape, and cloud extinction. These airborne data will be coupled with ground-based measurements to evaluate model simulations of Arctic climate. In addition, NASA

Cloud-Aerosol Lidar and Infrared Pathfinder Satellite Observations (CALIPSO) lidar and CloudSat radar are being combined with data from the A-train configuration of the Aqua, Aura, and Parasol satellites orbiting in formation to enable systematic observation of the key climate forcing of aerosol indirect effects, climate sensitivity of cloud feedbacks, and polar climate response of difficult-to-observe polar clouds. A number of long-term observing projects are now an integral part of AON, which comprises a total of 34 projects.

Argo Profiling Array. Argo profiling floats, measuring upper ocean temperature and salinity, have now been deployed in all oceans. The United States operates approximately half of the global array in cooperation with 22 countries operating the other half. The floats drift at depth and periodically rise to the sea surface, collecting data along the way, and report their observations in real-time via satellite communications. This global data set is used together with complementary data from satellites and other *in situ* systems to document ocean heat content and global sea-level change.

Atmospheric Radiation Measurement Program. The Atmospheric Radiation Measurement (ARM) Program is the largest global change research program supported by the U.S. Department of Energy (DOE). The primary goal of the ARM Program is to improve the treatment of cloud and radiation physics in global climate models in order to improve their climate simulation capabilities. To achieve this goal, ARM scientists and researchers around the world use continuous data obtained through the ARM Climate Research Facility. This scientific user facility provides a unique asset for interdisciplinary global change research among the national and international scientific communities. In addition, the ARM mobile facility (AMF) is used to study cloud and radiation processes in multiple climatic regimes. AMF can be deployed to sites around the world for durations of 6 to 18 months. Data streams produced by AMF are available to the atmospheric community for use in testing and improving parameterizations in global climate models. Using measurements from the ARM Mixed-Phase Arctic Cloud Experiment, a data set has been created that allows climate and cloud models to simulate Arctic weather, allowing for direct comparison of observations and model simulations.

Bermuda Atlantic Time-series Site (BATS) and Hawaiian Ocean Time-series (HOT). BATS and HOT are two oceanic sites for monitoring the carbon system of the ocean in response to climate change. BATS and HOT have been funded by NSF since 1988/1989. The primary research objectives of these ocean measurement programs are to establish and maintain deep-water hydrostations for observing and interpreting physical and biogeochemical variability, through repeat measurements of a suite of core parameters at regular intervals, followed by compilation of the data and rapid distribution to the scientific community. BATS and HOT have provided a baseline of consistent measurement and analysis of hydrographic and biological parameters throughout the water column, including a focus on carbon exchange between the oceans and atmosphere. Among other results, they have contributed substantially to the understanding of carbon removal pathways from the surface ocean. See <www.bios.edu/research/bats.html> and <hahana.soest.hawaii.edu/hot/hot_jgofs.html>.)

Climate Reference Network. The U.S. Climate Reference Network (USCRN) is a network of climate stations now being developed as part of a NOAA initiative. Its primary goal is to provide future long-term homogeneous observations of temperature and precipitation that can be coupled to long-term historical observations for the detection and attribution of present and future climate change. Data from USCRN are used in operational climate monitoring activities and for placing current climate anomalies into a historical perspective. USCRN will also provide the United States with a reference network that meets the requirements of the Global Climate Observing System (GCOS). If fully implemented, the network will consist of about 110 stations nationwide.

Forest Inventory and Analysis (FIA). As the Nation's continuous forest census, the FIA program projects how forests are likely to appear 10 to 50 years from now, to enable the evaluation of whether current forest management practices are sustainable in the long run and to assess whether current policies will allow the next generation to enjoy America's forests. FIA reports on status and trends in forest area and location; the species, size, and health of trees; total tree growth, mortality, and removals by harvest; wood production and utilization rates by various products; and forest land ownership. FIA is managed by the U.S. Department of Agriculture (USDA) Forest Service, and includes periodic measurements of forest area and location; species, size, and health of trees; forest land ownership; and soil, understory vegetation, tree crown conditions, coarse woody debris, and lichen community composition (on a sub-sample of plots), all of which are parameters of high significance to carbon inventory and cycling. See <www.fia.fs.fed.us/>.

Global Climate Observing System. GCOS is a long-term, user-driven system that integrates global networks placed strategically across the atmospheric , oceanic, and terrestrial domains, permitting better understanding of climate variability and change and supporting research toward improved understanding, modeling, and prediction of the climate system. GCOS addresses the total climate system including physical, chemical, and biological properties, and atmospheric, oceanic, terrestrial, hydrologic, and cryospheric components. GCOS builds upon, and works in partnership with, other existing and developing observing systems such as the WMO Global Observing System and Global Atmosphere Watch, the Global Ocean Observing System, and the Global Terrestrial Observing System. Many U.S. observation networks and platforms contribute to GCOS. It includes *in situ*, airborne, and space-based observational components.

Global Earth Observation System of Systems. The purpose of GEOSS is to achieve "comprehensive, coordinated, and sustained" observations of the Earth system in order to improve monitoring of the changing state of the planet, increase understanding of complex Earth processes, and enhance the prediction of the impacts of environmental change. GEOSS aims to enable all nations to benefit from access to timely, quantitative, and high-quality long-term global data and information as a basis for sound decisionmaking. GEOSS provides the overall conceptual and organizational framework to build toward integrated global Earth observations to meet user needs. It is a 'system of systems' consisting of existing and future Earth observation systems, supplementing but not supplanting the mandates and governance arrangements of those systems. The established Earth observation systems, through which many countries cooperate as members of the United Nations specialized agencies and programs and as contributors to international scientific programs, provide essential building blocks for GEOSS. The benefits to society cannot be achieved without data sharing. The success of GEOSS depends on data and information providers accepting and implementing a set of interoperability arrangements, based on non-proprietary standards with preference given to formal international standards.

Global Precipitation Measurement Mission. Motivated by the successes of the Tropical Rainfall Measuring Mission (TRMM) satellite and recognizing the need for a more comprehensive global precipitation measuring program, NASA and the Japan Aerospace Exploration Agency conceived a new Global Precipitation Measurement (GPM) Mission. A fundamental scientific goal of the GPM Mission is to make substantial improvements in global precipitation

observations, especially in terms of measurement accuracy, sampling frequency, spatial resolution, and coverage – thus extending TRMM's rainfall time series. To achieve this goal, the mission will consist of a constellation of low-elevation Earth-orbiting satellites carrying various passive and active microwave measuring instruments. The record of precipitation has been extended in recent years to include oceanic as well as land areas using satellite measurements from TRMM. This is an example of a key climate data set to be maintained and extended into the future. The GPM Mission will be used to address important issues central to improving the predictions of climate, weather, and hydrometeorological processes; to stimulate operational forecasting; and to underwrite an effective public outreach and education program, including near-real-time dissemination of televised regional and global rainfall maps.

GLOBALVIEW. GLOBALVIEW is designed to enhance the spatial and temporal distribution of atmospheric observations of CO_2 and methane. Measurement records from many international laboratories are integrated and extended to produce a globally consistent cooperative data product. GLOBALVIEW is specifically intended as a tool for use in carbon cycle modeling studies. The data product includes synchronized smoothed time series derived from continuous and discrete land surface, ship, aircraft, and tower observations; weight files; summaries of seasonal patterns, diurnal patterns (where relevant), sampling time-of-day (where available), and atmospheric variability; the derived marine boundary layer reference matrix used in the data extension process; uncertainty estimates; and extensive documentation. Time series measurements of CO_2 from surface-based global networks (including AmeriFlux and GLOBALVIEW) are used to document carbon fluxes at particular locations, drive models of regional and global carbon fluxes, and provide the basis for top-down inversion estimates of carbon sources and sinks. See www.cmdl.noaa.gov/ccgg/globalview/>.

Groundwater Climate Response Network. The USGS Water Resources Program, in cooperation with hundreds of Federal, state, and local agencies, collects nationally consistent information about the Nation's groundwater resources and helps define and manage those resources. The USGS maintains a network of wells to monitor the effects of droughts and other climate variability on groundwater levels. The Groundwater Climate Response Network consists of a national network of about 140 wells monitored as part of the Groundwater Resources

Program, supplemented by more than 200 wells monitored as part of the Cooperative Water Program that meet the same network criteria.

Historical Climatology Network. The U.S. Historical Climatology Network (USHCN) is a high-quality, moderate-sized data set of daily and monthly records of basic meteorological variables from over 1,000 observing stations across the 48 contiguous United States. Daily data include observations of maximum and minimum temperature, precipitation amount, snowfall amount, and snow depth from 1,062 stations; monthly data consist of monthly averaged maximum, minimum, and mean temperature and total monthly precipitation from 1,221 stations. Most of these stations are U.S. Cooperative Observing Network stations located generally in rural locations, while some are National Weather Service First Order stations that are often located in more urbanized environments. The USHCN has been developed over the years at NOAA's National Climatic Data Center (NCDC) to assist in the detection of regional climate change and to analyze U.S. climate.

Integrated Ocean Observing System. The Integrated Ocean Observing System (IOOS) is the U.S. coastal-observing component of the Global Ocean Observing System (GOOS) and is envisioned as a coordinated national and international network of observations, data management, and analyses that systematically acquires and disseminates data and information on past, present, and future states of the oceans. A coordinated IOOS effort is being established by NOAA via a national IOOS Program Office co-located with the <Ocean.US> consortium of offices consisting of NASA, NSF, NOAA, and the Navy. The IOOS observing subsystem employs both remote and *in situ* sensing. Remote sensing includes satellite-, aircraft-, and land-based sensors, power sources, and transmitters. *In situ* sensing includes platforms (ships, buoys, gliders, etc.), *in situ* sensors, power sources, sampling devices, laboratory-based measurements, and transmitters.

Integrated Ocean Drilling Program. The Integrated Ocean Drilling Program (IODP) is an international marine research program that explores the Earth's history and structure as recorded in seafloor sediments and rocks, and monitors sub-seafloor environments. IODP builds upon earlier successes of the Deep Sea Drilling Project and Ocean Drilling Program, which revolutionized our view of Earth history and global processes through ocean basin exploration. IODP greatly expands the reach of these previous programs by using multiple drilling platforms, including riser, riserless, and mission-specific,

to achieve its scientific goals. The first phase of the IODP produced high-resolution records of climate over past millennia from marine sediments. Recent examples include the first recovery of central Arctic Ocean marine sediment records and expeditions to drill methane hydrates, which may have been a cause of past abrupt climate change. Rapidly accumulating marine sediments remain the longest, most continuous record of past climate and environmental variability found on the planet.

International Polar Year Observations. A wide array of polar climate observations continues to be a CCSP focus as part of the International Polar Year (IPY). Combining orbital, airborne, ship-borne, and *in situ* measurements, the purpose of these observations is to provide the high-quality records needed to detect potential future climate changes in the cryosphere. IPY aims to leave a legacy of enhanced observational systems, facilities, and infrastructure, within the frameworks of existing and new long-term international research programs, including many CCSP activities. IPY observations and coordination activities include ocean observing systems in the Arctic and Southern Oceans, coordinated acquisition of satellite data produced from cooperating space agencies, and enhanced operational and community monitoring systems for astronomy, Sun-Earth physics, atmospheric chemistry, meteorology, biodiversity, permafrost, glaciers, and geophysics. IPY projects will collect a broad-ranging set of samples, data, and information, and share that data and information through new access and information tools. IPY efforts include engagement with northern communities as full partners in research, analysis, and assessment, and the provision of rigorous scientific information to decisionmakers.

Landsat Data Continuity Mission. The importance of Landsat data continuity was a priority for FY 2007 and was emphasized in the FY 2007 edition of *Our Changing Planet*. Landsat Data Continuity Mission planning continues toward a proposed 2010 launch. In October 2006, NASA and USGS announced the selection and research objectives for the Landsat Science Team. The Science Team will recommend strategies for the effective use of archived data from Landsat sensors and investigate the requirements for future sensors to meet the needs of Landsat users, including the needs of policymakers at all levels of government. The team will cooperate with other Earth-observing missions, both nationally and internationally. This will improve quantification of drivers and atmospheric forcings of climate change, contribute to improved projections of this change, and provide improved understanding of the present environment, its variability, and how it is changing.

Long-Term Ecological Research Sites (LTER). There are 26 LTER sites whose mission is to provide the scientific community, policymakers, and society with the knowledge and predictive understanding necessary to conserve, protect, and manage the Nation's ecosystems, their biodiversity, and the services they provide. See <www.lternet.edu>.

Moderate Resolution Imaging Spectroradiometer. The Moderate Resolution Imaging Spectroradiometer (MODIS) instrument has been operating successfully on NASA's Earth Observing System (EOS) Terra mission for over 6 years and on the Aqua mission for over 4 years. The MODIS instruments provide daily global observations of atmospheric, land, and ocean features with unprecedented detail. The capability of MODIS to observe global processes and trends is leading to better understanding of natural and anthropogenic effects on the Earth-atmosphere system, and to better performance of general circulation models. More than 100 'Direct-Broadcast' stations are now operating across the globe, enabling MODIS data to be obtained in near-real-time. About 800 user agencies or entities worldwide routinely use MODIS observations for regional applications.

National Phenology Network. The mission of the USA National Phenology Network (USA-NPN) is to facilitate systematic collection and free dissemination of phenological data from across the United States. This is primarily being done to support scientific research concerning interactions among plants, animals, and the lower atmosphere, especially the long-term impacts of climate change.

National Streamflow Information Program. The USGS National Streamflow Information Program (NSIP) is a nationally consistent streamgaging network with stable long-term monitoring sites and a rigorous program of data quality assurance, management, archiving, and synthesis. The five components of NSIP follow: (1) an enhanced nationwide base streamgage network; (2) intense data collection during floods and droughts, and additional analysis of these data; (3) periodic regional and national assessments of streamflow characteristics; (4) enhanced information delivery; and (5) methods development and research. NSIP produces multi-purpose, unbiased surface water information that is readily accessible to all users.

NOAA Cooperative Air Sampling Network. The NOAA cooperative air sampling network is an international effort to obtain discrete, weekly surface samples from NOAA baseline observatories, cooperative sampling sites,

and commercial ships. Samples, collected from more than 60 global sites, approximately 10 of which are in North America, are analyzed for CO_2, methane, carbon monoxide, molecular hydrogen, nitrous oxide, sulfur hexafluoride, and for stable isotopes of CO_2 and methane. Measurement data are used to identify long-term trends, seasonal variability, and spatial distributions of carbon cycle gases.

NOAA Tall Towers. The NOAA Tall Tower observational system was established to extend the monitoring of long-lived trace gases to continental areas. Variations of trace gases, especially CO_2, are largest near the ground, so using tall (>400 m) transmitter towers as platforms for *in situ* and flask sampling for atmospheric trace gases provides a direct link between ecosystem processes and the atmospheric imprint.

The Ocean Research Interactive Observatory Networks. The Ocean Research Interactive Observatory Networks (ORION) is a program that will focus and integrate the science, technology, education, and outreach of an emerging network of ocean observing systems. A major component of ORION is the Ocean Observatories Initiative. See <www.oceanleadership.org/ocean_observing>.

Ocean Surface Topography Mission. The accurate, climate-quality record of sea surface topography measurements – started in 1992 with the Ocean TOPography EXperiment (TOPEX)/Poseidon and continued in 2001 by the Jason satellite mission – will be extended with the Ocean Surface Topography Mission (OSTM). These missions have provided accurate estimates of regional sea-level change and of global sea-level rise. Ocean topography measurements from these missions have elucidated the role of tides in ocean mixing and maintaining deep-ocean circulation. Furthermore, quantitative determination of ocean heat storage from satellite measurements together with measurements from the global array of temperature/ salinity profiling floats known as Argo have confirmed climate model predictions of the Earth's energy imbalance that is primarily due to greenhouse gas forcing. The high levels of absolute accuracy and cross calibration make these missions uniquely suited for climate research. OSTM is a collaboration among NASA, NOAA, the French space agency CNES, and the European meteorological agency EUMETSAT, and has a planned 2008 launch.

Orbiting Carbon Observatory. The Orbiting Carbon Observatory (OCO) is a new mission, expected to launch in 2008, that will provide the first dedicated, space-based measurements of atmospheric CO_2 (total column) with the precision, resolution, and coverage needed to characterize carbon sources and sinks at regional scales and to quantify their variability. Analyses of OCO data will regularly produce precise global maps of CO_2 in the Earth's atmosphere that will enable more reliable projections of future changes in the abundance and distribution of atmospheric CO_2 and studies of the effect that these changes may have on the Earth's climate.

Soil Climate Analysis Network (SCAN) and Snowpack Telemetry (SNOTEL). SCAN is an in situ real-time meteor-burst telemetry observation network. SCAN and SNOTEL data are used in data assimilation for water supply forecasting; drought assessments; flood response; integrated pest management; land productivity in relation to soil moisture and temperature changes; and help in predicting shifts in wetlands, crop yields, and other ecosystems, and could serve as ground-truth for satellite soil moisture information. SCAN data are being examined for use in modeling carbon fluxes. At each SCAN site, minimum measurements include precipitation, relative humidity, wind speed, solar radiation, and barometric pressure above ground, and soil moisture and soil temperature at several depths below the surface. SNOTEL site measurements include snow depth and moisture content. See <www.wcc.nrcs.usda.gov>.

Solar Variability: Glory. The Glory mission will continue to be developed in FY 2008, and is planned to launch in 2009. It will carry a Total Irradiance Monitor (TIM) based on the Solar Radiation and Climate Experiment (SORCE) TIM design, with the same high-precision phase-sensitive detection capability. Glory will also carry an Aerosol Polarimeter Sensor (APS), which will improve ability to distinguish among aerosol types by measuring the polarization state of reflected sunlight. Both TIM and APS will provide key measurements of the minimum of Solar Cycle 24. This less-active portion of the 11-year solar cycle is especially crucial in estimating any long-term trends in solar output – a key to understanding the 20th-century context of global change, as the Sun is the single entirely 'external' forcing of the climate system that is unaffected by climate change itself.

Solar Radiation and Climate Experiment. The NASA SORCE mission has provided new insights into solar forcing, with SORCE making the most accurate measurements of the total solar irradiance (TSI) and the first daily observations of the solar spectral irradiance in the visible and near-infrared. During the past 11-year solar cycle, the Sun's increased brightness at solar maximum warmed the Earth's atmosphere by about 0.1°C in the 10 km nearest to the surface (the troposphere), 1°C near

50 km in altitude (at the top of the stratosphere), and 400°C at an altitude of 500 km in the thermosphere. Currently the Sun is at solar minimum, and the TSI observations indicate that the solar brightness has decreased by about 0.02% over the past decade. This result suggests that the solar forcing could reduce slightly the larger global warming trend due to human-induced greenhouse gases (e.g., burning of fossil fuels). Continued observations of solar irradiance, such as by SORCE, are key to understanding the long-term solar changes that provide a natural forcing to climate change.

Tropical Ocean-Global Atmosphere (TOGA) Tropical Atmosphere/Ocean (TAO) Array. The TOGA program, a 10-year study (1985-1994) of climate variability on seasonal to interannual time scales, included the accurate determination of basin-scale fluctuations in surface winds, sea surface temperature, upper ocean heat content, near-surface currents, and sea level in the tropical Pacific. Measurement of these oceanographic fields is required to describe fully the variability related to the El Niño-Southern Oscillation (ENSO), to understand the physical processes responsible for that variability, and to initialize and verify short-term climate prediction models. The TAO array (renamed the TAO/TRITON array on 1 January 2000) consists of approximately 70 moorings in the tropical Pacific Ocean, telemetering oceanographic and meteorological data to shore in real-time via the Argos satellite system. The array is a major component of the ENSO Observing System, GCOS, and GOOS.

U.S. Climate Variability and Predictability (USCLIVAR). NSF, along with NOAA, funds the USCLIVAR/CO_2 Repeat Hydrography program that since 2003 has measured the CO_2 system of the global ocean to monitor changing patterns of CO_2 in the ocean and provide the necessary data to support continuing model development that will lead to improved forecasting skill for oceans and global climate. See <ushydro.ucsd.edu>.

Selected Data Management Systems

Carbon Dioxide Information Analysis Center. DOE's Carbon Dioxide Information Analysis Center (CDIAC) provides comprehensive, long-term data management support, analysis, and information services to DOE's climate change research programs, the global climate research community, and the general public. The CDIAC data collection is designed to answer questions pertinent to both the present-day carbon budget and temporal changes in carbon sources and sinks. The data sets are designed to provide quantitative estimates of anthropogenic CO_2 emission rates, atmospheric concentration levels, land-atmosphere fluxes, ocean-atmosphere fluxes, and oceanic concentrations and inventories. CDIAC provides unrestricted, free distribution of its data products.

Distributed Active Archive Centers (DAACS). **Eight** NASA DAACs, representing a wide range of Earth science disciplines, comprise the data archival and distribution functions of the Earth Observing System Data and Information System (EOSDIS). DAACs are responsible for processing certain data products from instrument data, archiving and distributing NASA's Earth science data, and providing a full range of user support. There are more than 2,100 distinct data products archived at and distributed from the DAACs. These institutions are custodians of Earth science mission data until the data are moved to long-term archives. They ensure that data will be easily accessible to users. NASA and NOAA have initiated a pilot project to develop a prototype system for testing approaches for moving MODIS data into long-term NOAA archives. This pilot project is part of the evolution of the Comprehensive Large Array-data Stewardship System developed by NOAA. Acting in concert with their users, DAACs provide reliable, robust services to those whose needs may cross traditional discipline boundaries, while continuing to support the particular needs of their respective disciplines. DAACs serve a broad and growing user community.

Earth Observing System Data and Information System. NASA's EOSDIS provides convenient mechanisms for locating and accessing products of interest either electronically or via orders for data on media. EOSDIS facilitates collaborative science by providing sets of tools and capabilities such that investigators may provide access to special products (or research products) from their own computing facilities. EOSDIS has an operational EOS Data Gateway (EDG) that provides access to the data holdings at all the DAACs and participating data centers from other U.S. and international agencies. Currently, there are 14 EDGs around the world that permit users to access Earth science data archives, browse data holdings, select data products, and place data orders.

Global Change Master Directory. **The Global Change** Master Directory (GCMD) is an extensive directory of descriptive and spatial information about data sets relevant to global change research. The GCMD provides a comprehensive resource where a researcher, student, or interested individual can access sources of Earth science data and related tools and services. At present the GCMD database contains over 18,200 metadata descriptions of data sets from approximately 2,800 government agencies, research institutions, archives, and universities worldwide; updates are made at the rate of 900 descriptions per month. GCMD contains descriptions of data sets covering all disciplines that produce and use data to help understand our changing planet. GCMD includes metadata from disciplines including atmospheric science, oceanography, ecology, geology, hydrology, and human dimensions of climate change to facilitate multidisciplinary global change research (e.g., how climate change may affect human health). GCMD has made it easier for such data users to locate the information desired. A portal has been created in support of GEOSS. See <gcmd.nasa.gov>.

Earth Science Research, Education, and Applications Solutions Network. **Forty** Cooperative Agreement projects that are part of NASA's Earth Science Research, Education, and Applications Solutions Network (REASoN) have completed their first year. The REASoN projects are part of NASA's strategy to work with its partners to improve its existing data systems, guide the development and management of future data systems, and focus performance outcomes to further Earth science research objectives. In order to achieve these goals, the REASoN projects are organized to engage the science community and peer review process in the development of higher level science products; to use these products to advance Earth system research; to develop and demonstrate new technologies for data management and distribution; and to contribute to interagency efforts to improve the maintenance and accessibility of data and information systems.

Appendix 3: Relationship of CCSP Strategic Goals to GCRA Research Elements

The CCSP strategic goals serve two functions: (1) they articulate the most common questions regarding climate change to foster the development of research activities to address those questions; and (2) they serve as both a roadmap for and a touchstone against which progress is measured. Some common questions emerged from the FY 2008 edition of *Our Changing Planet* (CCSP, 2007b):

- To what extent and how is the climate system changing?

- What are the causes of these changes?

- What will the future climate be like and what effects will a changed climate have on ecosystems, society, and the economy?

- How can we best apply knowledge about ongoing and projected changes to decisionmaking?

CCSP's strategic goals have a direct relationship, by design, to the research elements outlined in the Global Change Research Act (GCRA) of 1990. The GCRA research elements are as follows:

- Global measurements, establishing and providing stewardship for the worldwide observations necessary to understand the physical, chemical, and biological processes responsible for changes in the Earth system on climate-relevant spatial and temporal scales

- Documentation of global change, including the development of mechanisms for recording changes that will actually occur in the Earth system over the coming decades

- Studies of earlier changes in the Earth system, using evidence from the geologic and fossil record

- Predictions, using quantitative models of the Earth system to identify and simulate global environmental processes and trends, and the regional implications of such processes and trends

- Focused research initiatives to understand the nature of and interaction among physical, chemical, biological, and social processes related to global change.

The GCRA research elements capture the major objectives and stages of scientific enquiry as they apply to the study of global change: observations and monitoring; development of time series data for historical and present climate; understanding of past climate and variability; understanding of the processes operating on Earth and how they respond to climate change; and forecasting future global change and its consequences. The CCSP strategic goals take this further. They encapsulate the GCRA elements and add the overarching strategic framework that is needed to ensure integration and coordination across the component research activities conducted by the participating agencies, and they add the crucial element of decision support. Table 1 depicts this robust approach to the long- and short-term conduct of global change research that is integrated, scientifically sound, and relevant to society's needs.

Relationship of GCRA 1990 Reearch Elements to CCSP 2003 Strategic Goals

Learn from the Past	Observe	Record	Understand	Predict	
		Scientific Process			
GCRA Studies of earlier changes in the Earth system, using evidence from the geologic and fossil record.	Global measurements, establishing and providing stewardship for the worldwide observations necessary to understand the physical, chemical, and biological processes responsible for changes in the Earth system on climate-relevant spatial and temporal scales.	Documentation of global change, including the development of mechanisms for recording changes that will actually occur in the Earth system over the coming decades.	Focused research initiatives to understand the nature of and interaction among physical, chemical, biological, and social processes related to global change.	Predictions, using quantitative models of the Earth system to identify and simulate global environmental processes and trends, and the regional implications of such processes and trends.	
CCSP Collect Information and Increase Scientific Knowledge CCSP Goal 1: Improve knowledge of the Earth's past and present climate and environment, including its natural variability, and improve understanding of the causes of observed variability and change.			Understand Causes CCSP Goal 2: Improve quantification of the forces bringing about changes in the Earth's climate and related systems.	Characterize Uncertainties CCSP Goal 3: Reduce uncertainty in projections of how the Earth's climate and related systems may change in the future.	Understand Sensitivities and Predict Responses CCSP Goal 4: Understand the sensitivity and adaptability of different natural and managed ecosystems and human systems to climate and related global changes.
					Support Decisionmaking CCSP Goal 5: Explore the uses and identify the limits of evolving knowledge to manage risks and opportunities related to climate variability and change.
		Growth In Knowledge			

www.ingramcontent.com/pod-product-compliance
Lightning Source LLC
Chambersburg PA
CBHW080643180526
45168CB00008B/3291